OPTICAL DEVICES IN COMMUNICATION AND COMPUTATION

OPTICAL DEVICES IN COMMUNICATION AND COMPUTATION

Editor

Brian Ismay

scitus
academics

Optical Devices in Communication and Computation
Edited by **Brian Ismay**

ISBN: 978-1-68117-597-3
Library of Congress Control Number: 2016946452

www.scitusacademics.com

Notice

Printed in United States of America on Acid Free Paper

PREFACE

Optical communication and computing devices have a major influence on our daily lives, although we may not be aware of their existence, as in the case of intercontinental fiber cables connecting people worldwide, making it a worldwide community. Fiber optic communication is a way of transmitting information from one area to another by transmitting light pulses via an optical fiber. Light forms an electromagnetic carrier wave to carry information. Fiber optics, originally produced in the 1970s, revolutionized the telecommunications industry and played an important role in creating the information age. The advantages of optical fibers over electrical transmission have largely replaced copper wire communications in the advanced world's core networks.

This compendium emphasizes the recent developments in theoretical analysis, the design of new nano-photonic structures and functional materials for optical instruments. Thin optical fibers carry vast amounts of information from one place to another everywhere on this planet. There are many desirable characteristics of optical fibers to carry this information. They have enormous capacity, low cost information and have immunity from many disturbances that can affect electrical wires and wireless communication links. The superiority of optical fibers for the transport of information from place to place leads to a rapid replacement of older technologies. Optical fibers have played an important role in enabling the remarkable growth in worldwide communications that has taken place over the past 25 years and are essential for the proliferation of the Internet.

TABLE OF CONTENTS

1

NOVEL OPTICAL DEVICE MATERIALS – MOLECULAR-LEVEL HYBRIDIZATION

Kyung M. Choi [1]

[1] University of California at Irvine, USA

1. INTRODUCTION

Silicate glass has been widely used for optical device materials due to its excellent optical transparency. To satisfy our multiple demands in advanced optical device materials, organic/inorganic hybrid composites have been widely prepared by a bulk mixing technique, which is physically mixing multiple components at the bulk scales.

However, conventional glassy materials have shown limitations to modify their physicochemical properties by inserting desired components into glassy hosts. In addition, a significant phase separation occurs during a mixing process of multiple components, especially immiscible phases.

To overcome those limitations, there are growing interests in doping organic components or semiconductor particles into glassy hosts without any phase separation to combine beneficial properties at the molecular scales, and thus to bring desired properties (Figure 1). [1]

Hybrid materials lie at the interface of the organic and inorganic material regimes, where versatility in molecular tailoring approach offers novel molecular modifications in design of new chemical structures. Hybrid materials can also range, depending on the method of formation and domain size, from physical mixtures of inorganic oxides and organics (blends, composites) to nanocomposites and molecular composites that utilize formal chemical linkages between the organic and inorganic domains on the molecular scale.

Hybrid materials are ranged from the bulk-scales to molecular scales as shown in Figure 2 to mix up multiple components. [1-10]

Usually, hybrid materials mixed at the bulk scales retain the original properties of the individual organic and inorganic components. In other words, their final properties are significantly influenced by the characteristics and their domain sizes of individual components after the mixing process at the bulk scales.

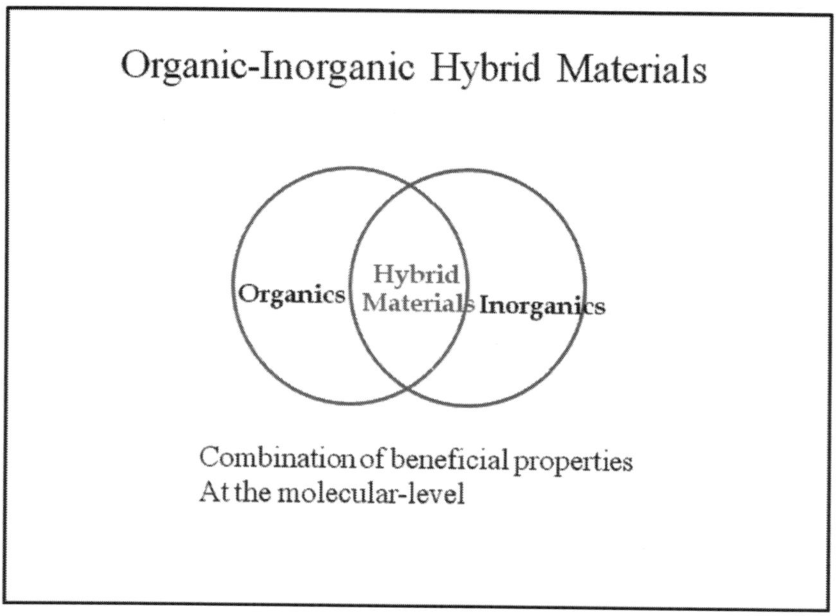

Figure 1. Organic/inorganic hybrid materials. [**1**]

In addition, a phase separation problem also limits to achieve a uniform mixing of multiple components during the bulk-mixing process.

To overcome those limitations of the bulk-scale mixing technique, *molecular-level hybridization technique* has been actively investigated. [1-10] This technique results in molecular-level composites, which are their domain sizes in the nanometer scale often create new properties, which would not be expected from those individuals by the loss of individuals' identities after the molecular-level mixing process thereby creating new properties.

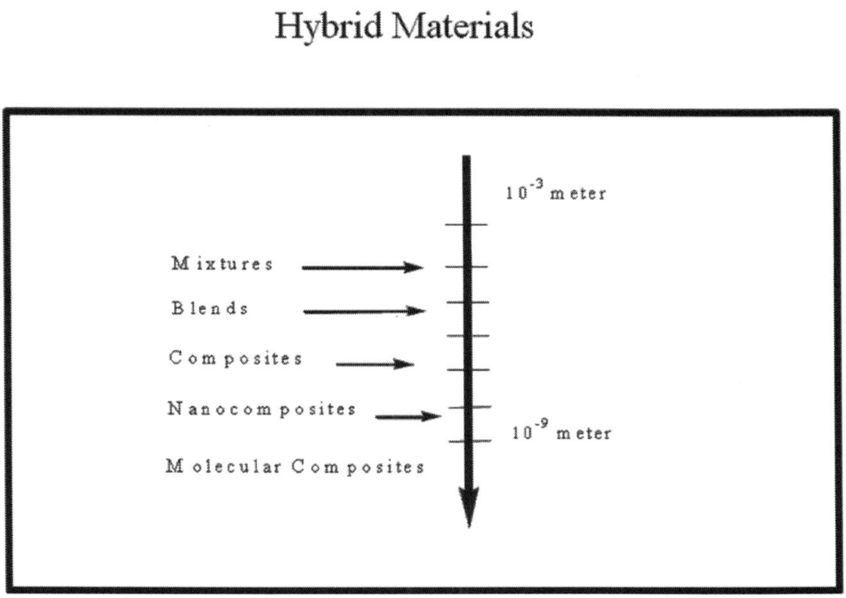

Figure 2. Relative size scale of mixing domains for different types of hybrid materials. [**1-10**]

Especially for optical device materials, desired properties of hybrid glasses can be chemically designed and then prepared by incorporating functional organic fragments between inorganic oxides.

With this strategy, novel optical device materials with beneficial properties can be obtained by embedding organic spacers into silicate network to create organically modified hybrid glasses as demonstrated in earlier publications (Figure 3). [1-10] Furthermore, the molecular-level mixing technique doesn't show any significant phase separation; because, *the molecular-level hybridization* is based on a microscopically homogeneous mixing and thus the uniform distribution of organic and inorganic moieties in a domain size at the molecular level is provided.

Those organically modified glasses mixed at the molecular scale, were provided by a molecular modification, which inserts desired organic fragments between two inorganic oxides to create entirely new optical properties. Figure 3 shows *the molecular-level hybridization* to produce polysilsesquioxanes.

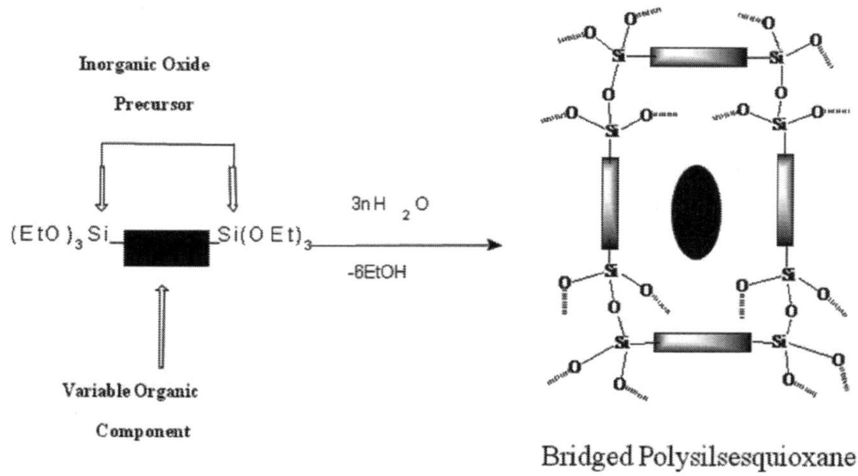

Figure 3. Molecular-level hybridization technique to produce polysilsesquioxanes. [**6**]

Bridged polysilsesquioxanes are a new family of molecular-level composites, which also a new version of hybrid glasses; these are prepared from hybrid sol–gel processable monomers through a sol-gel polymerization. These are microscopically homogeneous with uniform distribution of the organic spacers and thus also show an excellent optical transparency. There are many publications of novel optical device materials based on organic/inorganic hybrid glasses. [11-17]

In optical device materials, organically modified silicate glasses/transparent polymers have been actively pursued to develop novel optical devices, such as lasers, optical switches, optical fibers, waveguides, laser amplifiers, optical displays, and data storage devices.

[11-17]

In addition, we can also control the porosity of those organically modified silica, polysilsesquioxanes, by inserting different molecules or sizes of organic spacers as shown the void space in Figure 3 (right); due to the insertion of organic spacers, the porosity of the resulting hybrid glass significantly increased. [18-26]

The expanded pores allow us to dope semiconductor or metal particles without any significant mechanical cracks. *The molecular-level hybridization* also solves the phase separation problem during molecular-level mixing process.

The pore size of organically modified silicate materials can be controlled by both choices of organic spacers and sol-gel conditions. Those molecularly designed hybrid glasses have shown high surface areas and a relatively narrow distribution of pore sizes that range from the high micropore to the low mesopore domain (15-100 Å).

Several organic/inorganic hybrid sol-gel monomers have been molecularly designed and then synthesized for developing novel hybrid glasses. A variety of bridged polysilsesquioxanes, organically modified hybrid silica, have been designed by *the molecular-level hybridization* (Figure 4). [18-26]

Those functional organic spacers inserted in the silicate networks (Figures 3 and 4) can serve as dopant precursors to growth particles in the porous glassy hosts. Those void volumes in the glassy hosts can be a matrix for the growth of quantum dots, such as semiconductor or metal particles. No phase separation occurs during the corporations of those dopants.

In Figure 5, those sol-gel processable monomers contain functional groups, which are sol-gel polymerized under either acidic or base condition, and then produce highly porous xerogels. [4]

Due to the highly porous silicate matrices, we doped various dopants without any phase separation; for example, we prepared highly nano-porous polysilsesquioxane systems, and then controlled sol-gel conditions to dope nano-sized transition metal particles or semiconductor particles, such as CdS [18,19, 21], chromium [20, 21], iron [22], cobalt [27], and platinum [28] into the silicate hosts.

2. NOVEL OPTICAL DEVICE MATERIALS FOR LASER AMPLIFIER [25]

We have developed novel laser devices based on those organic/ inorganic hybrid glasses prepared by*the molecular-level hybridization.*

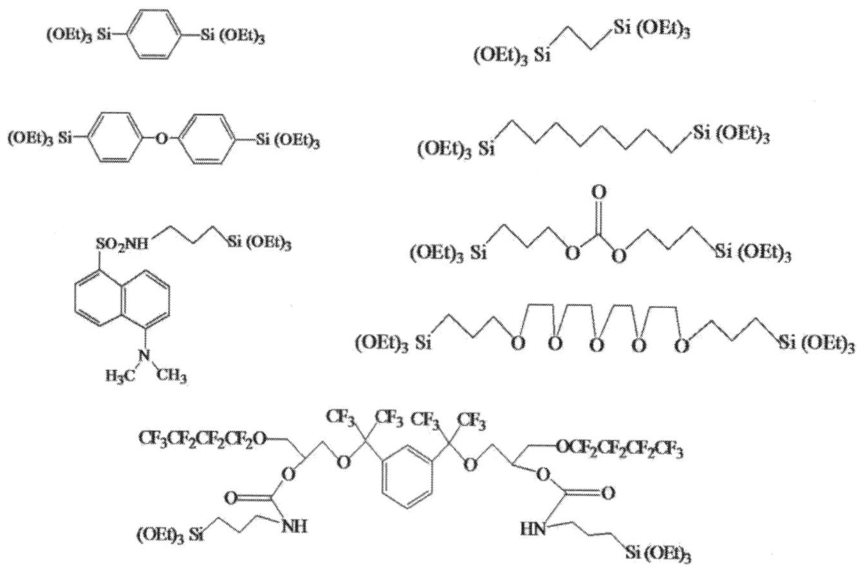

Figure 4. A variety of sol-gel processable monomers prepared by *the molecular-level hybridization*.

2.1. Rare-Earth Ion Doped Laser Amplifier

Silicate-based optical fibers and planar waveguide amplifiers are widely studied for optoelectronic applications because of the superior chemical resistance and compatibility with other optical devices based on polymeric materials.

Transparent silica doped with rare-earth metal ions has been used for laser amplifiers, such as photonic fiber amplifiers, solid-state lasers, compact laser amplifiers, ultra short pulse lasers, high-power lasers, so on. [29-39]

Planar waveguides and fibers doped with rare-earth metal ions are a key challenge; thus, an enormous amount of publications in rare-earth metal ion doped optical devices has been found. [29-37] Fabrication of optical devices with high resolution offers an efficient approach to minimize the cost and size of optical amplifiers.

High gain laser amplifiers can be achieved by improving several relative factors, such as optical losses, phonon energies, pumping powers and distances, fluorescence life times, and refractive indices of optical medium. The lasing efficiency can be also improved by changing the chemical environment of rare-earth metal ions.

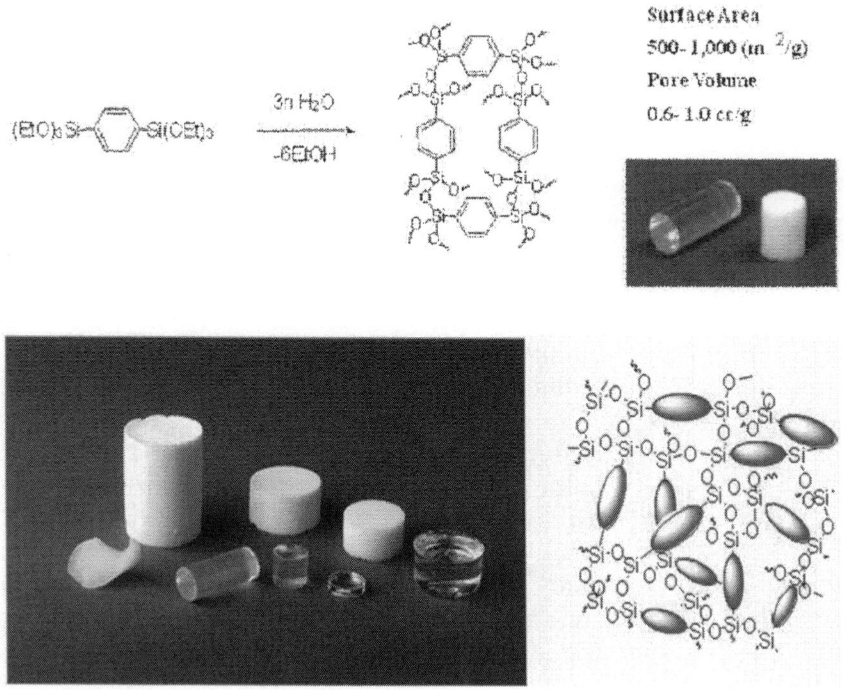

Figure 5. Highly porous polysilsesquioxanes. [**4**]

In a laser amplification based on rare-earth metal ions, erbium (Er) has been widely used as a gain medium due to its strong fluorescence at 1540 nm, which is a useful wavelength in optical amplifications. Especially, erbium-doped fiber amplifiers (EDFA) dominate this object of high gain optical amplifications. [32-35]

Since the performance of laser amplifiers is significantly influenced by optical media, scientists have been investigating organic/inorganic hybrid silicate hosts doped with Er^{3+} ions, in order to achieve high NIR efficiency and low phonon energy of the matrix to shorten the pumping distance and thus to obtain proper gain/life time. [32-35, 38, 39]

However, inorganic based optical media have shown a limitation to adjust the chemical environment of doped metal ions. For example, conventional silicate-based laser amplifiers often fail to produce high lasing performance because of a strong absorption raised from the OH-group at 1540 nm. The low concentration of erbium-ions in silicate host and the small absorption cross-section of the erbium-ions also limit the performance of normal silicate-based laser amplifiers since the doping level of rare-earth ions in glassy hosts significantly depends upon lasing efficiency.

Organic/inorganic hybrid glasses have been actively pursued as an alternative to conventional silicate glass for fabricating laser devices due to low temperature process and the promise of bringing new optical properties that are not possible from inorganic silica. [1, 40-42]

Furthermore, organic/inorganic hybrid silica is a good candidate to adjust the fluorescence environment of rare-earth metal ions by incorporating desired organic precursors into glassy hosts without any phase separation; usually, in inorganic silicate hosts, rare-earth ions tend to aggregate due to the absence of non-bridging oxygens, which cause a significant deduction of lasing efficiency. The synthesis of chemically modified silica with homogeneous doping of rare-earth ions is a key contributor to improve the performance of laser amplifications.

Optical device materials are required good optical transparency, controllable porosity, chemical purity, tunable refractive index, so on. Our goal for achieving those desired properties is to improve the fluorescence environment of Er^{3+} -ions in glassy hosts. For that, we devoted our attention to achieve an excellent chemical homogeneity of Er^{3+}-ion environment in glassy hosts.

2.2. Design Of Fluoroalkylene-Bridged Xerogel Doped With Er $^{3+}$/Cdse

To demonstrate enhanced performance in laser amplifiers, we designed fluoroalkylene-bridged polysilsesquioxanes doped with Er^{3+}/CdSe nano-particles. [25] The fluoroalkylene-bridged silica was initially designed to reduce the phonon energy of the glassy host. Furthermore, CdSe nano-particles were also provided for further manipulation of the photochemical environment of erbium-ions in the matrix.

Use of organic silanes incorporates an organic fragment as an internal component of the silicate network. Sol–gel polymerization involves the hydrolysis of ethoxysilyl groups to yield silanols follows by subsequent condensation to form siloxane (Si–O–Si) linkages. In the sol-state, the condensation is insufficient to form a network, and the solution remains processable.

When sufficient cross-linking occurs, a network is formed and the transition from the sol-state to the gel state occurs. The presence of organic fragments within the 3-D structure imparts organic character to the hybrid glass that changes the microenvironment of additional fluorescence rare-earth ions incorporated in the glassy host.

In this study, fluorocarbon-linkages were designed to achieve high hydrophobicity within the hybrid glassy matrix. Erbium isopropoxide was also employed as the source for the Er^{3+} ions. Furthermore, CdSe nano-particles were also prepared and incorporated into the fluorinated glassy matrix to reduce the phonon energy of the glassy host (the phonon energy of CdSe = 200 cm^{-1}). [43]

In principle, when rare-earth metal ions are excited in transparent glassy matrices, they can behave as a laser, which enables amplification of the incident light intensity. The lasing performance significantly relies on rare-earth metal ion doping level, host materials' physicochemical property, and chemical homogeneity.

2.3. Experimental

A set of three different sol–gel processable monomers were prepared; tetraethoxy-silane (TEOS), 1,6-bis (triethoxysilyl)hexane, and 2,2,3,3,4,4,5,5-octafluoro-1,6-hexanediol bis(3-triethoxysilyl)propyl carbamate.

2.3.1. Tetraethoxysilane (Teos)

TEOS was purified by drying over 4 Å molecular sieves followed by a vacuum distillation.

2.3.2. Synthesis Of 1,6-Bis(Triethoxysilyl)Hexane

1,6-Bis(triethoxysilyl)hexane was synthesized by a 'hydrosilylation' of the corresponding α, ω-alkyldienes with triethoxysilane employing chloroplatinic acid (H_2PtCl_6) as a catalyst. 1,5-Hexadiene (12.3 g, 0.15

mol), triethoxysilane (54.1 g, 0.33 mol), and chloroplatinic acid (1mL of 7.5×10^{-5} mol in isopropanol) were placed in a round bottle flask. After 10 hours of simple stirring process at a room temperature, the reaction mixture darkened.

The reaction was monitored by GC. The crude product was purified by a vacuum distillation with a resulting purity of 99.87 % by GC analysis: bp 130 °C/0.1 mmHg. The final product was verified by NMR analysis and mass spectroscopic analysis and was consistent with data previously report. [11]

2.3.3. Synthesis Of 2,2,3,3,4,4,5,5-Octafluoro-1,6-Hexanediol Bis(3-Triethoxysilyl)Propyl Carbamate

2,2,3,3,4,4,5,5-Octafluoro-1,6-hexanediol (1g, 3.8 mmol) and 3-isocyanatopropyl triethoxy-silane (1.9 g, 4.3 mmol) were placed in a round bottom flask with a magnetic stirrer. The flask was sealed, purged with nitrogen, and 10 mg of dibutyl-tin-dilaurate was injected into the vial using a syringe as a catalyst.

The reaction was kept at room temperature under a nitrogen flow for several hours and monitored for the disappearance of the isocyanate peak at 2270 cm^{-1} (CN) in FT-IR. As the isocyanate group was converted to urethane group, identical peaks of 3400 cm^{-1} (NH) and 1720 cm^{-1} (CO) were observed. The product was dissolved in methanol and the tin catalyst was completely removed using a separated funnel. Methanol was then removed by a rotary evaporator. The product was used without further purification.

2.3.4. Er^{3+}-Ion/Cdse Doping Procedure

A sol-gel processible monomer, erbium isopropoxide (Chemat Technology), was used as a source of Er^{3+}-ions. CdSe nano-particles were synthesized by a previously reported procedure. [44, 45] Those dopants were incorporated by mixing with the appropriate sol–gel mixtures (Table 1); xerogels-a5 and -a10 denote higher erbium concentrations (5 and 10 times higher) than that of xerogel-a system.

We also examined the homogeneity of three sol–gel mixtures (Figure 6). Ethanol was used as a solvent. A visual inspection was carried out to determine the comparative homogeneity of those sol–gel mixtures containing erbium isopropoxide (Figure 6-1) or both of erbium isopropoxide/CdSe nano-particles (Figure 6-2). In a mixing test (Figure 6), erbium isopropoxide was indicated as a bright pink color in the photographs. The CdSe nano-particles also show a characteristic bright orange-color in those mixtures.

Table 1. List of xerogels doped with different components.

Xerogels	Silicate matrices	Dopants
T-Xerogel	TEOS	None
H-Xerogel	Hexylene-	None
F-Xerogel	Fluoroalkylene-	None
Xerogel-a	TEOS	Erbium ions
Xerogel-b	TEOS	Erbium ions/CdSe
H-Xerogel-a	Hexylene-	Erbium ions
H-Xerogel-b	Hexylene-	Erbium ions/CdSe
F-Xerogels-a, -a5, and -a10	Fluoroalkylene-	Erbium ions
F-Xerogels-b, -b5, and -b10	Fluoroalkylene-	Erbium ions/CdSe

2.3.5. Sol–Gel Procedures

Those sol–gel mixtures were then polymerized to produce the condensed xerogels under acidic condition using HCl as a sol-gel catalyst. Those xerogels were kept in a vacuum oven for 1–2 days to remove the remaining solvent and complete condensation.

2.3.6. Solid State Nmr Experiments

The ^{29}Si solid state NMR analysis of condensed xerogels was performed using a Varian Unity 400 solid state NMR spectrometer. The degree of condensation of undoped xerogels was computed from the single pulse magic angle spinning (SP/MAS) technique. A line fitting routine was also used in the analysis of the ^{29}Si NMR resonance in each spectrum to establish the siloxane ratio in the different structures.

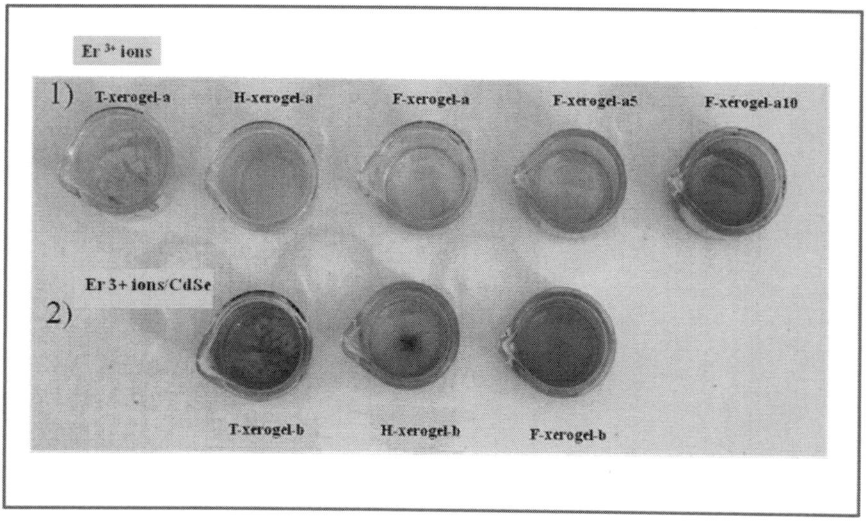

Figure 6. Mixing test.

2.3.7. Fluorescence Measurements

Fluorescence study of Er^{3+}-ions doped into those glassy matrices was carried out by using the Ar+ ion laser (488 nm). Laser power densities ranging from 1.5 to 3 Wcm^{-2} were used for the measurements.

2.4. Results and Discussions

2.4.1. Mixing Test

During the mixing test (Figure 6), we observed that the TEOS-based mixtures revealed a substantial degree of undesirable phase separations after doping with the Er^{+3}-ion sources or Er^{+3}/CdSe nano-particles. For example, in Figure 6-1, the T-xerogel-a mixture shows a significant phase separation even at the lower erbium concentration.

In contrast, hybrid sol–gel monomer mixtures showed significantly less phase separations (Figure 6). Hybrid sol–gel monomer mixtures accommodate and homogeneously distribute the Er^{3+}/CdSe source without any significant phase separation. In mixing test with hybrid sol-gel monomer mixtures, we observed no momentous phase separation, especially in the highly fluorinated sol-gel mixtures.

It is apparent that the TEOS-based sol–gel mixture has a rather limited solubility of erbium-ions, and hence a limited capability for fluorescence enhancement.

Subsequently, we provided CdSe nano-particles by following the earlier method [44, 45], and then added CdSe nano-particles into those sol–gel monomer mixtures containing the the Er^{3+}/CdSe source. Usually, CdSe nano-particles synthesized in colloidal configuration are suitable for incorporation into a variety of hosts including sol–gel mixtures. The comparative homogeneity of the three sol-gel monomer mixtures containing both of the erbium isopropoxide and CdSe nano-particles is also shown in Figure 6-2.

In Figure 6-2, TEOS-based mixture (T-xerogel-b) shows the CdSe nano-particles segregated in the mixture; the orange-colored CdSe nano-particles are observed to phase separate within the mixture.

In contrast, a hybrid sol-gel monomer system (H-xerogel-b) shown in Figure 6-2, the CdSe nano-particles mixed better than the T-xerogel-b. In the H-xerogel-b mixture, most of CdSe particles were dissolved, except some of undissolved orange-colored CdSe residues toward the middle of the container. In fluorinated mixture (F-xerogel-b) shown in Figure 6-2, CdSe nano-particles were incorporated without phase separation. This result demonstrates that the fluoroalkelene-bridged sol-gel monomer has the capability of uniformly incorporating both types of dopants without any phase separation.

2.4.2. Solid State Nmr Analysis

The chemical composition and the degree of condensation for those condensed xerogels can be determined by solid state nuclear magnetic resonance, infrared, and Raman spectroscopies. [1]

We employed a solid state NMR analysis to determine the degree of condensation of hybrid glassy hosts. ^{29} Si solid state NMR was used to identify the Si–O–Si bonds in variety states of condensation for three matrices. Single pulse magic angle spinning NMR methods were employed for the characterization of T-xerogel, H-xerogel, and F-xerogel to calculate the degree of condensation.

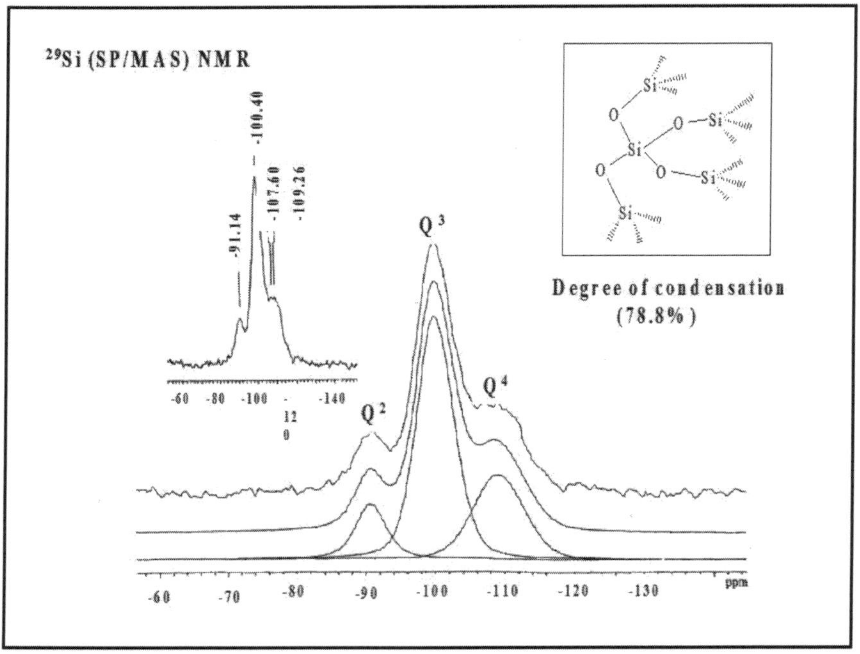

Figure 7. Si solid state NMR spectrum for undoped T-xerogel.

Figures 7-9 show the result of ^{29}Si SP/MAS solid state NMR analyses and peak deconvolution lines of both normal and modified silicate systems.

Figure 7 shows the ^{29}Si SP/MAS solid state NMR spectrum of the undoped T-xerogel. As shown inFigure 7, it reveals three peaks, which correspond to the Q^2, Q^3, and Q^4. The degree of condensation for the T-xerogel was calculated to be 78.8%.

Figure 8 shows the ^{29}Si SP/MAS solid state NMR spectrum of the undoped H-xerogel. Three peaks shown in spectrum correspond to the T^1, T^2, and T^3. The degree of condensation of the H-xerogel was also computed to be 79.1%.

The ^{29}Si SP/MAS solid state NMR spectrum of the undoped F-xerogel is given in Figure 9. The T^1peak is not observed in this system. Also, the T^3 peak intensity in Figure 9 is higher than that of T^2. We found that the degree of condensation was dramatically increased to 91.1% in this case.

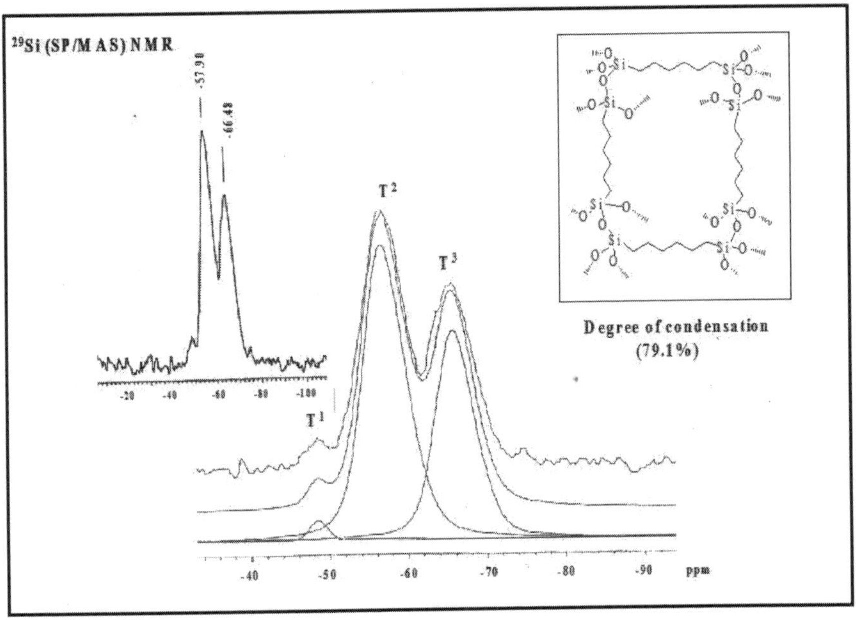

Figure 8. Si solid state NMR spectrum for undoped H-xerogel.

The high degree of condensation in the undoped F-xerogel can be explained as a result of the electron-withdrawing effect of the fluorine, which causes a partial positive charge at the silicon facilitating nucleophilic attack in the sol–gel process thus accelerating hydrolysis and condensation.

The high degree of condensation, disappearance of T^1 peak, and enhanced T^3 peak in the F-xerogel indicate a low level of hydroxyl group content and a greater degree of condensation when fluorinated alkylene groups are present in the glassy hosts.

Si–OH groups show a high phonon energy (3000–3500 cm^{-1}) at 1540 nm. [43]

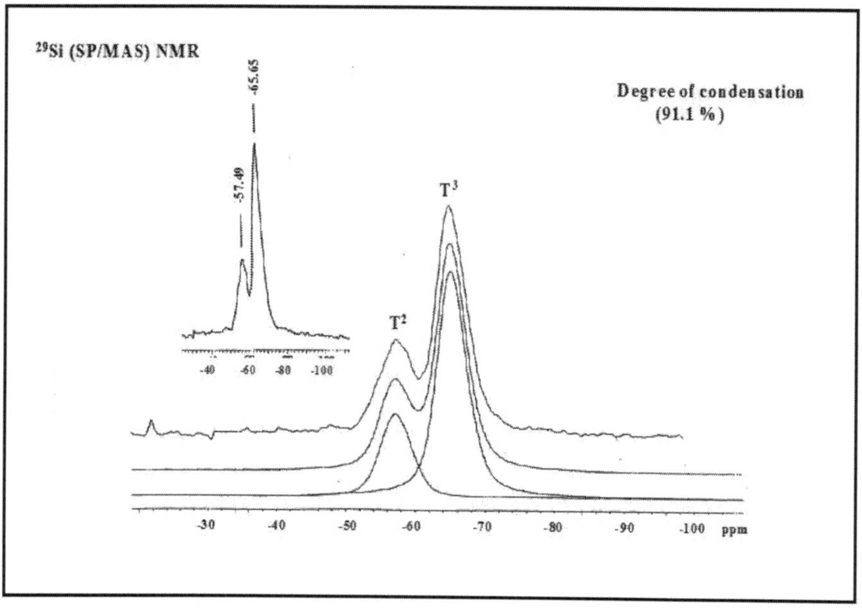

Figure 9. Si solid state NMR spectrum for undoped F-xerogel.

The reduction in hydroxyl content in the F-xerogel matrix decreases the phonon energy of the matrix. The exclusion of moisture from the high hydrophobicity of the fluoro-alkylene groups in the F-xerogel matrix may also contribute to reduce the absorption at 1540 nm with a concomitant increase the fluorescence intensity of Er^{3+}-ions.

All these effects contribute to the increased fluorescence from erbium-ions in the fluorinated hybrid glassy matrix.

2.4.3. Xps Analysis

We also employed a XPS study to determine the chemical composition of F-xerogel-a. [46]

A full XPS scan was obtained in the 0–1100 eV range. Detail scan was also recorded for the Er (4d) region. Figure 10 shows a full spectrum for F-xerogel-a. The XPS spectrum of F-xerogel-a shows an erbium peak at ~169 eV, which corresponds to the presence of Er_2O_3.

The atomic compositions were evaluated in this study. The concentration of each element (atomic %) was calculated; O(1s)—22.74 atomic %, C(1s)—53.29 atomic %, Er(4d)—1.49 atomic %, F(1s)—9.02 atomic %, Si(2p)—10.91 atomic %, N(1s)—2.55 atomic %. From the XPS analysis, it was estimated that the F-xerogel-a contains ~1.49 atomic % of erbium.

2.4.4. Fluorescence Measurements

Erbium-ions incorporated into glassy matrices exhibit well defined energy level transitions in 4f-shell electronic configurations.

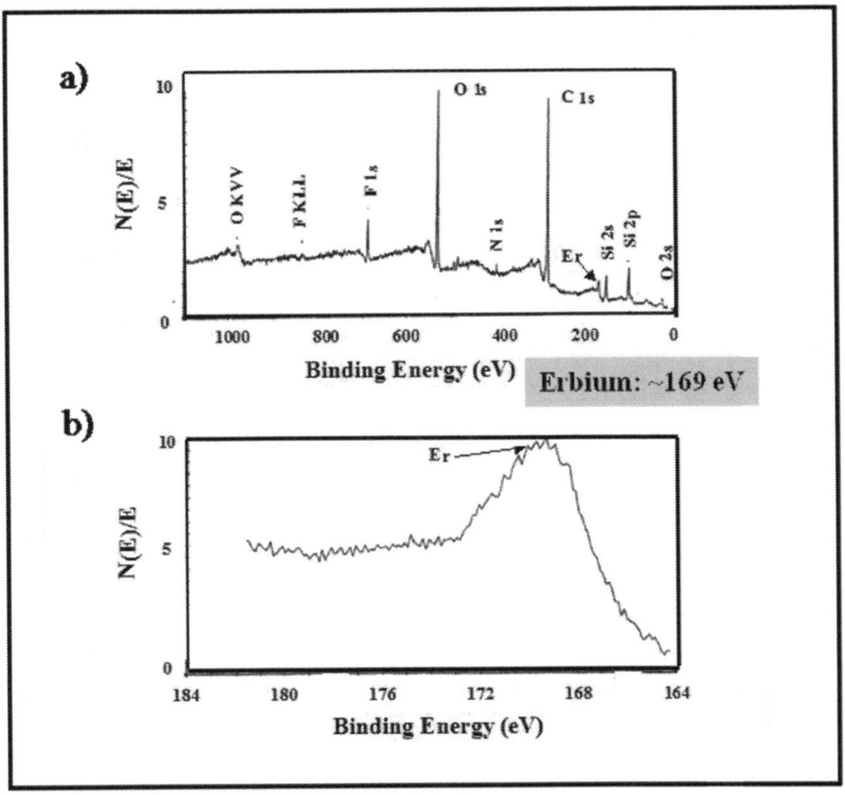

Figure 10. XPS Analysis of F-xerogel-a.

For erbium-ions, the $4I_{13/2}$ to $4I_{15/2}$ transition is important in optical communications; because, it results in fluorescence at 1540 nm, which is the most important wavelength regime for optical communication applications, especially in long-distance telecommunication networks. [47]

We examined how the fluorescence intensity of erbium-ions was dependent upon the matrix environment when fluorine and CdSe nano-particles were incorporated into hybrid glassy hosts. We carried out fluorescence analysis of erbium-ions around 1540 nm. The results are shown in Figures 11-13.

A comparison of fluorescence intensities from erbium-ions doped into different glassy hosts is shown in Figure 11 (H-xerogel-a and F-xerogel-a).

In Figure 11, fluorescence intensity of erbium-ions increased significantly more in F-xerogel-a than the other hybrid system of H-xerogel-a.

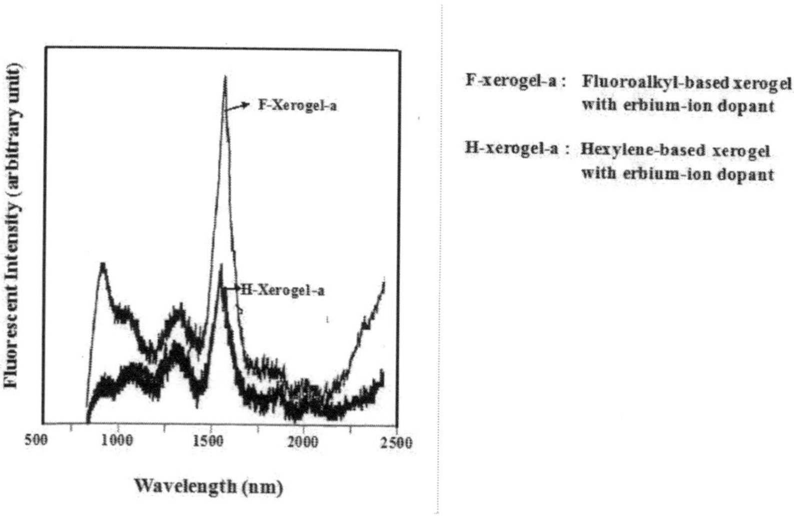

Figure 11. Photoluminescence of erbium-ions; different silicate matrices. Comparative fluorescence intensities of two different hybrid xerogels doped with Er^{+3} ions using a low power of 1.5 Wcm^{-2}.

The explanation of Figure 11 is that the enhanced fluorescence intensity of the fluorinated matrix is attributed to mainly its high hydrophobicity combined with the lower OH-group contents, which revealed in NMR study in Figure 9. Solid state silicon NMR analysis indicates an enhanced condensation in fluorinated xerogel compared to that of alkylene-bridged xerogel. The fluorinated hosts also showed an excellent chemical homogeneity in mixing test (Figure 6), which significantly affects the lasing performance.

Further investigations in fluorescence studies have been carried out. Since it is important to consider the erbium-ion concentration effect, fluoroalkylene-based glasses doped with two different levels of erbium concentrations were prepared for the determination of erbium-concentration effect (Figure 12).

Intensity of those xerogels (F-xerogel-a and F-xerogel-a5) was measured at a power density of 3 Wcm^{-2}. As shown in Figure 12, the fluorescence intensity increases (the upper curve) in the case of F-xerogel-a. It can be explained that the higher erbium concentration of F-xerogel-a5 caused a low lasing efficiency due to the "self-quenching effect."

Fluoroalkylene-bridged xerogel containing Er^{3+}-ions shows significantly reduced absorptions at the 1540 nm by reducing amounts of uncondensed hydroxyl groups.

The presence of CdSe nano-particles also significantly influences the fluorescence environment of Er^{3+}-ions in different glassy hosts, resulting in the increased fluorescence intensity. [43]

CdSe nano-particles were used to modify the photochemical environments of erbium-ions in glassy hosts by taking advantage of a low phonon energy of CdSe phase (200 cm^{-1}) [43], since the incorporation of semiconductor nano-particles resulted in an enhancement of the semiconductor-to-erbium transfer when the quantum well and erbium-ion transition energies became close.

We thus examined the fluorescence of erbium-ions surrounded by CdSe nano-particles since it was anticipated that the presence of CdSe in the modified glassy hosts would affect the fluorescence performance (Figure 13). In order to test this, we prepared fluoroalkylene-bridged hybrid glasses doped without and with the CdSe nano-particles, F-xerogels-a5 and F-xerogel-b5, respectively.

As shown in Figure 13, the fluorescence intensity of F-xerogel-b5 is dramatically increased, which indicates an improvement in lasing efficiency by modifying photochemical environments of erbium-ions.

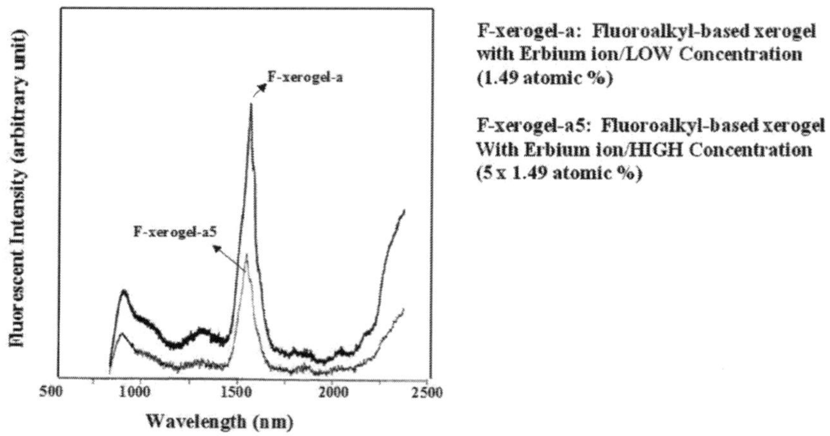

Figure 12. Photoluminescence of erbium-ions; different erbium concentrations. Comparative fluorescence intensities of xerogels with different Er^{3+}-ion concentrations (F-xerogels-a and -a5) using a power of 3.0 Wcm^{-2}.

By taking advantage of the structural features and uniform doping capability in modified glassy matrices, we successfully demonstrate that the fluorescence environments of erbium-ions can be a key to improving the performances of optical devices like laser amplifiers to overcome the limitations in inorganic silica.

In conclusion, we have demonstrated here a promising result in laser amplifications by employing bridged polysilsesquioxanes doped with Er^{3+}-ions/CdSe nanoparticles.

To avoid the high phonon energy raised from OH-groups (3000–5300 cm^{-1}) at 1540 nm [43], we designed highly fluorinated hybrid glasses to shorten the fluorescence-level life times of dopants, which adversely affect optical device performance.

The presence of CdSe nano-particles, by virtue of its lowering of phonon energy, also appears to significantly influence the nature of the surrounding photochemical environment of Er^{3+}-ions in the fluorescence study.

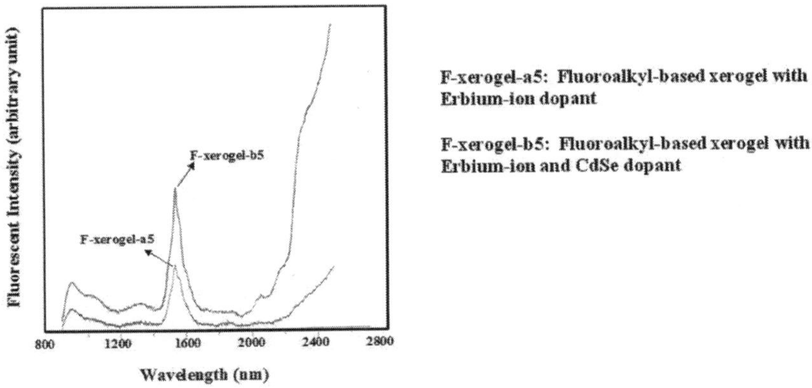

Figure 13. Photoluminescence of erbium-ions; CdSe nanoparticle effect. Comparative fluorescence intensities of fluoroalkylene-bridged xerogels doped without and with the CdSe nano-particles (F-xerogels-a5 and -b5), respectively, using a high power of $3Wcm^{-2}$.

From those study, we found that the control of such optical materials' properties affect the performance of optical devices via molecular tailoring strategy, which is molecular-level hybridization technique.

3. NOVEL OPTICAL DEVICE MATERIALS FOR ACOUSTIC WAVE [26]

3.1. Novel Optical Device Materials Based On Polysilsesquioxanes

The preparation of semiconductors, metals, and ions in a variety of transparent materials has been actively pursued for optical devices. Scientists have taken a great attention to incorporate metal particles/ions in glassy hosts to develop high performance optical devices.

Chemists also have taken intensive challenges on how to achieve homogeneous incorporations of semiconductors or metal particles in glassy hosts, which are significantly influenced by diffusion of reagents, the number of nucleation sites, stabilization of growing particles by surface functionality, the boundary constraints of the growth matrix, and the opportunity for equilibration or "ripening" of particles formed under kinetic growth conditions. [2, 21, 48, 49]

Organically modified hybrid glasses, which were prepared by *'molecular-level hybridization'* (Figure 3), are a good candidate to develop laser device materials due to their easy processability and chemical modification.

Polysilsesquioxanes can be prepared by *'molecular-level hybridization'*, which inserts different types and sizes of organic spacers between two inorganic oxides. The sol-gel chemistry was employed to covalently incorporate semiconductors/metals in various oxidation states as an integral component of hybrid silicate matrices.

The 'molecular-level corporation techniques' also can be employed for the incorporation of semiconductors or transition metals/ions dispersed in optically transparent silica to prepare novel optical devices.

Especially, polysilsesquioxanes are microscopically homogeneous due to uniform distribution of organic and inorganic moieties in a domain size at the molecular level. Hence, these hybrid glasses can be used for optical device materials since the molecular-scale mixing process significantly reduces phase separation and thus produces high quality of optical clearance.

In this work, we molecularly designed a novel hybrid glass to develop alkylene-bridged polysilsesquioxanes doped with Cr^{o}/CrOx phases (Figure 14). The doped xerogel film effectively generates a HUGE ACOUSTIC WAVE.

3.2. Periodic Alignment Of Alkylene-Spacers To Create Molecular-Scale Optical Grating For Acoustic Wave

Usually, scientists prepare 'periodic metal frames' to create 'diffraction grating' at the bulk-scales. For example, in a spectroscopic monochromator, 'optical grating effect' is used to split, and then diffract the light into several beams traveling in different directions.

In 'grating equation', the directions of these beams rely on the spacing/distances of the grating, which has a periodic structure of the media, and the wavelength of the light. Gratings, which modulate the phase rather than the amplitude of the incident light, can be also produced. A key idea of the periodic alignment of alkylene-spacers is to create 'a molecular-scale diffraction grating effect' and thus to generate a huge acoustic wave.

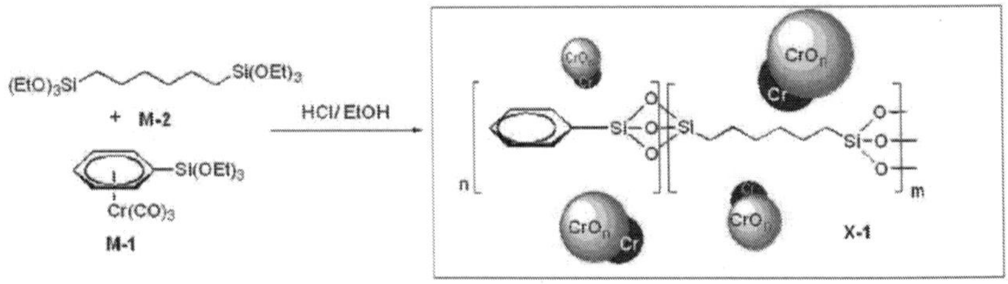

Figure 14. Hexylene-bridged polysilsesquioxane doped with Cr°/CrOx.

Figure 15 illustrates a light that passes through 'optical grating at the molecular-scales' designed in the hybrid silicate host. As you can see, the periodic alkylene-spacers create 'effective diffraction grating at the molecular scales' and thus produces a huge acoustic wave. This is a new optical phenomenon, which hitherto hasn't been discovered.

In addition, the distance between those alkyl-spacers can be also "controlled" by both choices of different organic spacers and sizes of sol-gel monomers. This is an effective method of developing new laser device materials based on organically modified silica, polysilsesquioxanes.

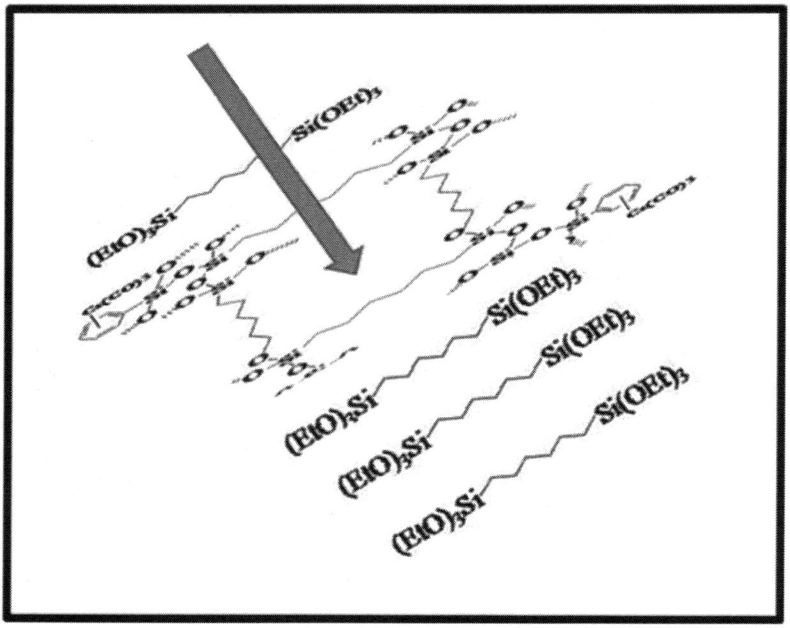

Figure 15. Periodic alignment of alkyl-chains to create "optical grating" at the molecular scales.

Usually, when the laser beam goes through a solid medium, the density wave is linear; because, in solid media, heat doesn't decay through the solid medium effectively.

Interestingly, the highly compressed xerogel, hexylene-bridged polysilsesquioxane doped with Cr°/CrOx phases showed a strong 'acoustic response' as much as a liquid.

In our laser experiments, we calculated physical parameters of hexylene-bridged xerogel doped with Cr°/CrOx phases. The coefficient of phonon diffraction (D) of the doped xerogel was FIVE times smaller than that of normal glass. Which means that the thermal conductivity of the doped xerogel, is FIVE times less than that of normal glass.

In addition, the diffraction efficiency, absorption light efficiency, (45%) of the doped xerogel is higher than that of methanol (25%), which means the COMPRESSIBILITY of the doped xerogel is as effective as liquid. The acoustic refractive intensity and acoustic response generated from the doped xerogel was as strong as liquid. Therefore, the doped xerogel serves as a 'heat generator,' and thus the heat gets transferred into expansion or compression wave (acoustic waves) effectively.

There were a lot of efforts to develop novel laser device materials based on organic and inorganic hybrid silica. [50-54] Especially, this is a unique approach to create an effective optical grating, which brings a HUGE ACOUSTIC RESPONSE by setting up a molecular-scale optical grating effect. By building up the periodic structure of long alkyl-chains in the hybrid glassy hosts, we obtained a HUGE ACOUSTIC WAVE, which hasn't been found. This is a new optical phenomenon.

3.3. Novel Sol-Gel Condition To Produce Thin Films Of Alkylene-Bridged Xerogel Doped With Cr^o/Crox

The new optical property partially rose from a new sol-gel mixing condition, which effectively produces HIGHLY COMPRESSED, THIN xerogel films. Conventional sol-gel conditions often result in poor optical transparency/mechanical property. For example, xerogels obtained from the conventional sol-gel conditions are thick and brittle materials. Those thick xerogels are difficult to handle, especially for optical applications.

The limitation motivated us to find a novel sol-gel mixing condition, which produces a highly compressed, thin sol-gel film with a smooth surface and excellent optical clarity.

We discovered a novel sol-gel mixing condition, which results in such a thin xerogel film of alkylene-bridged polysilsesquioxanes doped with Cr^o/CrOx with a low thermal conductivity and high compressibility (Figure 16). [26]

During the sol-gel polymerization, a volume of the sol-gel mixture was dramatically reduced, and then left a green sol-gel film with an excellent optical clarity (Figure 16).

For a source of chromium, we have synthesized a sol-gel processable chromium precursor [20, 26]; we prepared a green sol-gel film based on hexylene-bridged polysilsesquioxanes doped with Cr^o/CrOx phases using the chromium precursor. [20, 26]

In Figure 16 (top left), it shows a undoped hexylene-bridged polysilsesquioxane prepared from our novel sol-gel mixing condition; it was plastic-like a xerogel monolith. As you see in Figure 16 (top, left), it shows an excellent optical transparency.

Doped xerogels were also prepared in Figure 16 (top, Right and bottom). Those green xerogels were obtained as "plastic-like thin films" with an excellent optical compressibility and low thermal conductivity. It was prepared without any mechanical damage/cracking problem.

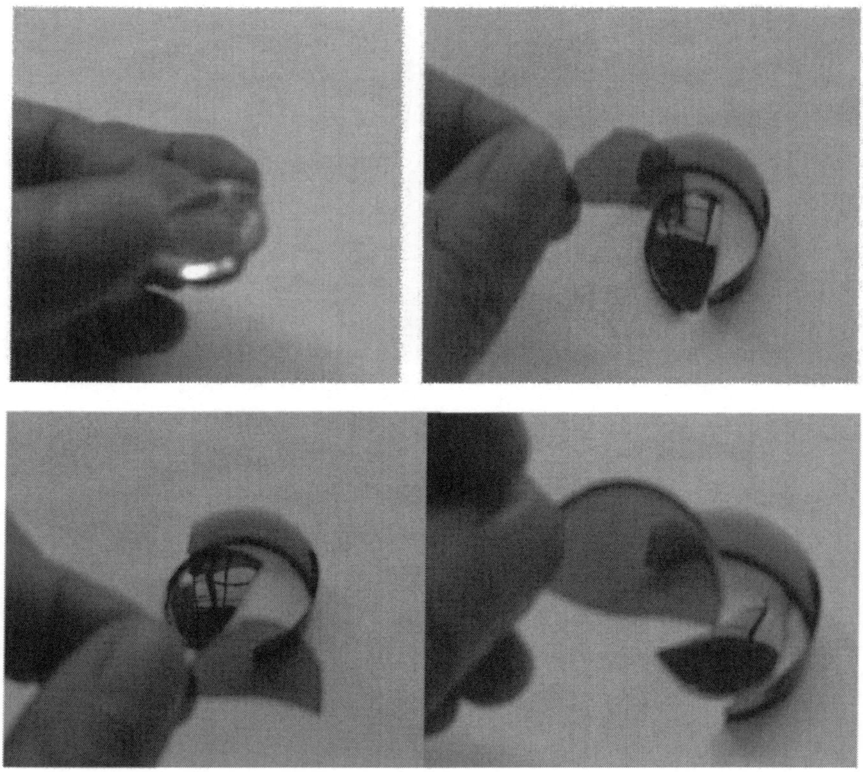

Figure 16. A comparative photos of undoped- (top left) and the Cr^{o}/CrOx doped- (top right and bottom) xerogels based on hexylene-bridged polysilsesquioxane under an acidic condition.

3.4. Experiments

3.4.1. Preparation Of Doped Xerogels

Syntheses of sol-gel monomers, 1,6-bis(triethoxsilyl)hexane and chromium precursor, η^6-chromium tricarbonyl(triethoxysilyl)benzene, were carried out by following the earlier procedures, respectively. [11, 20]

Subsequently, hexylene-bridged xerogel doped with Cr^o/CrOx phases was prepared by a copolymerization of η^6-chromium tri-carbonyl (triethoxysilyl)benzene and 1,6-bis(triethox-silyl)hexane under an acidic condition (Figure 17). A green glass was provided after the sol-gel polymerization (Figure 16).

Density of the doped xerogel was measured to be 1.2696 g/cm^3 using a He-gas pycnometer. From AAS analysis, the chromium amount was also analyzed to be 1.4% by weight.

3.4.2. Tem, Edax, and Electron Diffraction Analysis

A novel periodic alignment of alkylene-spacers in hybrid glass was specifically designed for creating 'a molecular-scale diffraction grating.' We employed TEM analysis to identify the molecular alignment built up in the hybrid silicate matrix.

The doped hybrid glass under an acidic condition (Figure 17) was ground to powders with a particle size (<150 μm), and deposited on a plasma-etched carbon substrate supported copper grid. TEM dark-field images were obtained using a Philips transmission electron microscope (TEM: CM 20/STEM, PW 6060).

The energy-dispersive X-ray diffraction (EDAX) pattern of the glassy particles was also obtained by an EDAX analyzer (Philips TEM-EDAX, PV 9800). For the electron diffraction, the camera length was calibrated experimentally with a gold standard, and an X-ray spectrum analyzer at 200 kV was used.

3.4.3. Laser Analysis

To establish the optical properties of doped xerogels, we prepared a sample (<1mm thickness) fabricated with η^6-chromium tricarbonyl (triethoxysilyl)benzene loading of 2.4 % under an acidic sol-gel condition. The thin xerogel film exhibited a nonlinear property. The nonlinear optical (NLO) properties of doped xerogel films were measured by the degenerated into four wave mixing (DFWM) technique. [55] A quartz sample was used as a reference.

We used two types of lasers, a YAG laser with 50 ps pulse-width at 532 nm and a dye laser with 150 fs pulse-width at 640 nm. The output of either of lasers is split into two strong pump beams and a weak probe. The delay between two pump pulses is set to zero to create interference patterns in the doped sol-gel film. Variable delay line on the probe beam allows to measure the decay time of the diffracted beam.

3.5. Results and Discussions

Figure 17 describes a sol-gel procedure for the preparations of hexylene-bridged polysilsesquioxanes doped with Cr^{o}/ CrOx.

The sol-gel process was carried out by a copolymerization of two sol-gel monomers, η6-chromium tricarbonyl(tri-ethoxysilyl)benzene (M-1) and 1,6-bis(triethoxsilyl)hexane (M-2) (Figure 14).

Those sol-gel monomers can be produced bridged Si-O-Si network under either acidic or basic condition, and thus processed into transparent glasses, glassy films, fibers, xerogels, and aerogels, and monoliths. [1,6] From the basic condition, it produced hybrid glasses containing chromium metal particles; because, the M-1 was stable under the basic sol-gel condition.

In contrast, the acid-catalyzed system produced thinner sol-gel films than those of glasses obtained under base condition (Figure 16). Under the acidic sol-gel condition, decomposition of the M-1 competes with sol-gel copolymerization. The product of acid catalyzed decomposition reaction is "chromium oxides" and H_2 (Figure 17). [56-58]

In Figure 18, TEM images of alkylene-bridged silica doped with Cr^{o}/CrOx phases reveal unusual nano-fringe patterns, which rose from the lattice fringes of the aligned alkyl-spacers in the silicate matrix. The novel molecular design results in 'a molecular level grating characteristic' when laser light passes by those periodic carbon-chains (Figure 15).

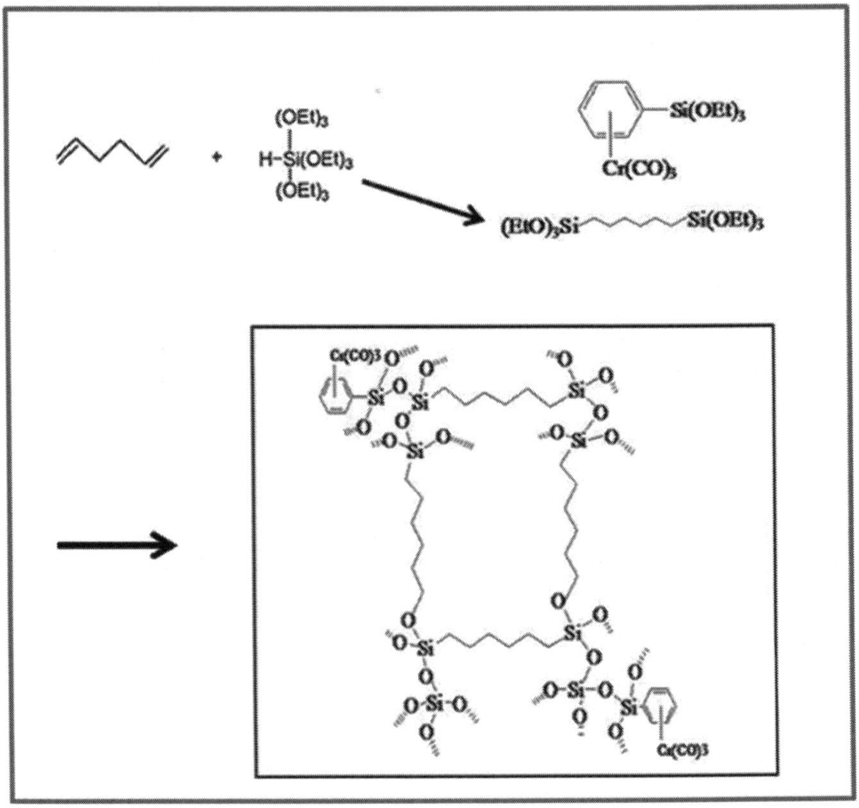

Figure 17. Synthesis of hexylene-bridged xerogel attached with the Cr^{o} precursors.

Based on those nano-fringe patterns, an effective optical grating was created in the hybrid silicate matrix. The TEM images of the doped hybrid glass reveal nano-fringe patterns, which are highly organized nano-periodicity (pointed with arrows in Figure 18). The nano-periodic patterns are sustained over substantial domains and appear to arise from lattice fringes.

In short, the formation mechanism of these nanoperiodic features observed in the TEM images arises from the highly arranged alkylene-spacers in the sol-gel monomer (Figure 15).

EDAX and electron diffraction analyses of these dark regions shown in the TEM images were also performed. In the EDAX spectrum, a Cr (Kα) peak was observed; thus, the dark contrast shown in the TEM images

(Figure 18) was identified as a chromium phase spread over the hybrid glassy host by both of EDAX/electron diffraction analyses.

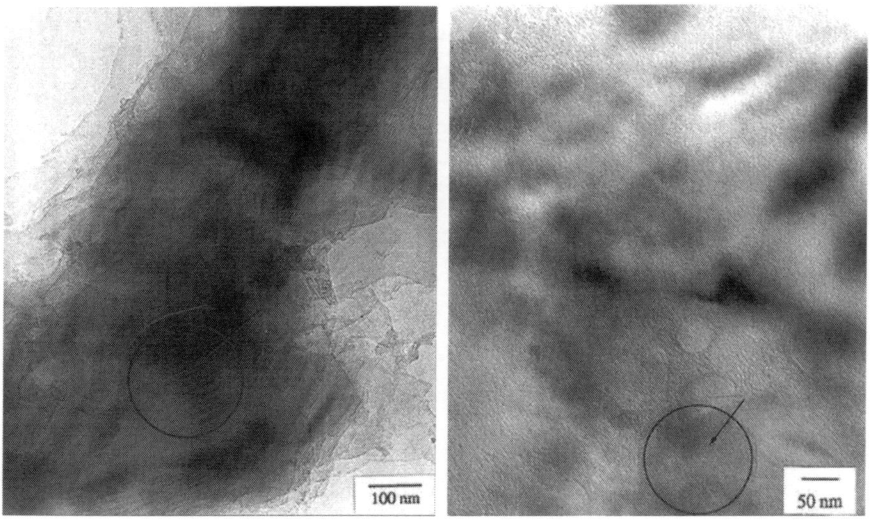

(a) (b)

Figure 18. a and b. TEM Image of Cr°/CrOx-doped hexylene-bridged polysilsesquioxane under acidic condition; it reveals unusual nano-fringe patterns with alternating features of a lattice spacing of about 50 Å in different areas.

The Cr°/CrOx phases were produced by a chemical reaction, which is a simultaneous sol-gel copolymerization then decomposition of the chromium precursor under the acidic condition. The electron diffraction pattern of the dark areas in TEM images was also identified as a mixture of Cr°/ CrOx phases.

In order to calculate a lattice space of those highly organized nano-fringe patterns shown in the TEM images, we also carried out electron diffraction analysis of the doped sol-gel film. The result is shown in Figure 19. The nano-periodicity gives rise to features in the electron diffraction pattern.

Figure 19 shows the circled diffraction patterns that a rise from crystalline chromium metal and a set of diffractions near the center of the beam corresponded to the nanofringes observed in TEM images (Figure 18).

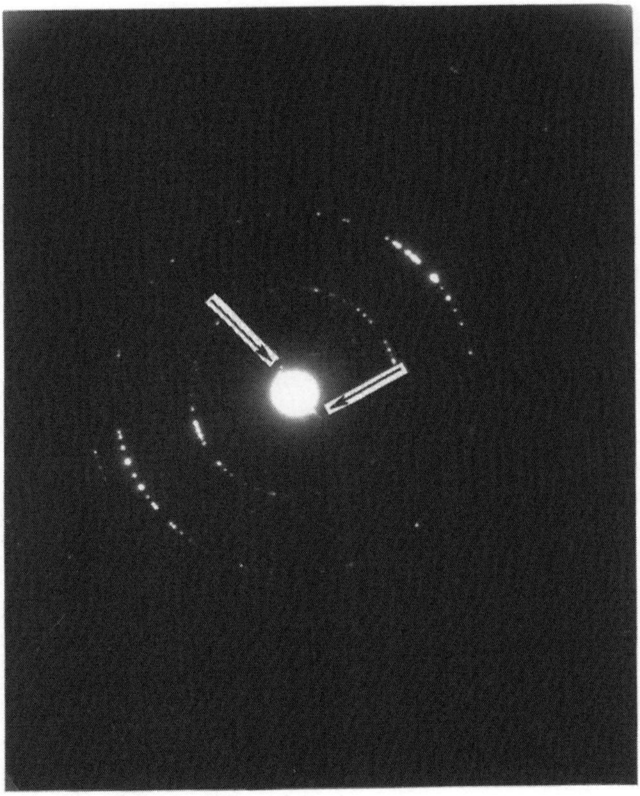

Figure 19. Electron diffraction pattern of doped hybrid glass; circled diffraction patterns correspond to the d-spacings of chromium metal. Highlighted features (arrows) correspond to the nanoperiodicity in the TEM images in **Figure 18**. From a distance between two diffraction spots point out each arrow a lattice spacing of the nanoscale fringes was calculated to be ~50 Å.

From the diffraction pattern corresponded to the nano-fringes, a lattice space of the nanostriped patterns observed in TEM images was calculated about 50 Å from a distance between two diffraction spots in two sets of diffraction patterns, which are pointed with arrows in Figure 19; each set consists of two diffraction spots.

The distance between alkyl-spacers significantly relies on the optical grating efficiency, and thus it can be "controlled" by inserting different types and sizes of organic spacers.

As shown in Figure 14, the reactions occurring during the formation of the sol-gel xerogels are complex and include simultaneous sol-gel copolymerization and decomposition of M-1 under the acidic condition. At present, the mechanism of formation of these nanoperiodic features observed in the TEM images may arise from the highly arranged alkyl-spacers in the sol-gel monomer (M-2).

In this study, the nonlinear optical (NLO) properties of the doped xerogel film were measured by the degenerated four wave mixing (DFWM) technique. [12, 13, 55]

In femto- and pico-second experiments, "electronic χ^3" and "population χ^3" of the doped xerogel film have been measured (Figure 20). The DFWM signals for both "pure electronic and population χ^3" are shown in Figure 20 as a small spike around t = 0. It is asymmetric and longer than the laser pulse. The trailing edge of the peak has decay, which is probably connected with population relaxation.

In thermal nonlinearity, the coefficient of phonon diffraction, which is proportional to the coefficient of thermal conductivity, has been calculated from the DFWM experiments as 1.9×10^{-3} cm^2 /sec using an equation, $D = \Lambda^2/4\pi\tau$ (where, Λ is the period of the diffraction grating and τ is the decay time of the thermal signal).

This number is about FIVE times lower than that of a normal glass. In other words, the thermal conductivity of the doped xerogel is FIVE times less than that of a normal glass. In acoustic study, the doped xerogel also shows an interesting new optical property, *an effective generation of a large acoustic wave.*

When the temperature of the doped xerogel at the maximum of the interference pattern goes up, the material expands then a counter propagating wave of expansion and compression start traveling inside the glassy host. Since the index of refraction depends on density of the material, on top of slow decaying thermal grating, we will have *a dynamic diffraction grating propagating with the sound velocity.*

By changing the delay on the laser probe beam, we measured *the period of acoustic grating and extract the sound velocity of the material* (Figure 20). We used YAG laser at 532 nm and 50 ps pulse-width for a laser analysis for the doped xerogel obtained by the novel sol-gel condition, which produces thin xerogel films with unusually high compressibility (higher density).

Small spike around t= 0 corresponds to the signal due to electronic nonlinearlity (Figure 20). The time required for the acoustic wave to travel from one interference maximum to another is twice the time between t=0 and the peak of the acoustic signal. The sound velocity (C) in the thin doped xerogel was calculated from the distance between two acoustic waves (Δτ) as C = 3.2 x 10 ^{5} cm/sec.

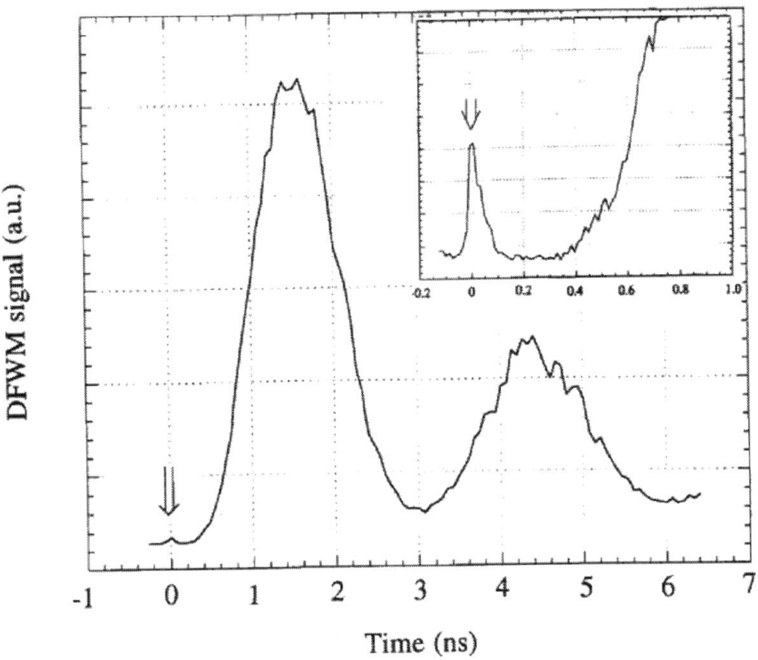

Figure 20. DFWM signals obtained from the doped xerogel measured in femto-second experiment (in a box) and pico-second experiment.

The signal decay time raised from the doped xerogel film was evaluated by the continuous wave (CW) probe experiment (Figure 21); it was obtained as 17 μsec.

Therefore, the doped xerogel serves as a heat generator in the slow nonlinearities due to the low thermal conductivity and high compressibility of the hybrid glass, thus the heat is transferred into expansion or compression wave (acoustic wave) very efficiently.

We also measured the diffraction efficiency of the probe beam at the delay time, corresponding to the peak of acoustic signal. At energy level about P = 0.47 J/cm^2, which is close to the optical damage threshold, the diffraction efficiency was 45 %.

The amplitude of the acoustic signal will depend upon how effectively the laser pulse energy is transferred to an expansion wave, which in turn depends on compressibility of the host materials. For a comparison we did the same measurement for a dye solution in methanol with the same optical density and the same energy density. The diffraction efficiency of methanol was 25 %, which means the compressibility of the doped xerogel film is as effective as liquid.

In a conclusion from the laser experiments, the doped xerogel film has a lower thermal conductivity than that of a normal glass. The compressibility of the doped xerogel is sufficiently high so the density grating formed in the doped xerogel could be effective to create high diffraction beam in the sound velocity.

We also believe that the nano-fringe patterns revealed in the TEM images (Figure 18), which rose from the lattice fringes of alkylene-bridged silicate matrix may result in an effective grating characteristics when the light passes by those long carbon-chains of alkylene-based polymeric networks.

The characteristics observed from the novel doped xerogel are new photonic properties, which hitherto have not been possible from simple physical mixing process of individual components at the bulk scale. In other words, those new optical properties are created by *the molecular-level hybridization*.

Based on these experiments, the ($Cr^o/CrOx$)-doped polysilsesquioxane with low thermal conductivity and high compressibility are suggested as a new type of optical device materials for optical applications, for example diffraction beam modulators.

We demonstrated here some examples of creating new optical properties by designing novel molecular building blocks at the molecular-scales via *the molecular-level hybridization.*

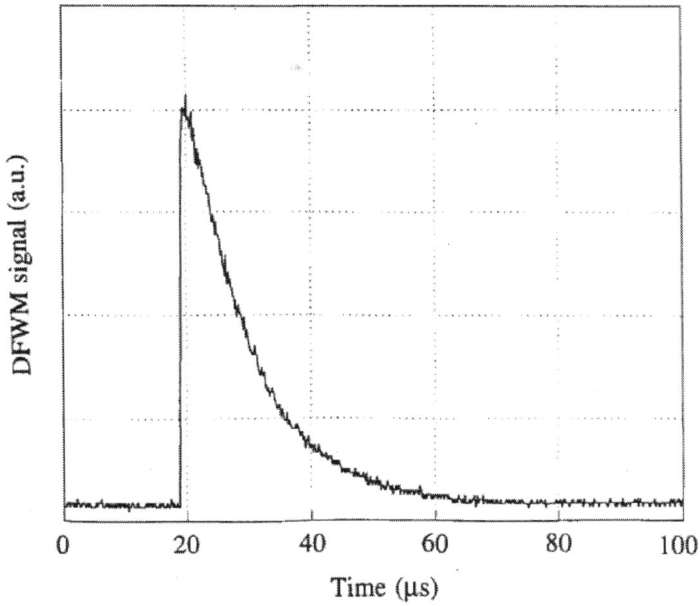

Figure 21. The decay of thermal signal for the doped xerogel measured in CW probe experiment.

4. CONCLUSIONS

We introduce the *molecular-level hybridization* for the preparation of polysilsesquioxanes, which are hybrids of inorganic oxides and organic network polymers.

Optically transparent hybrid glasses are prepared by a molecular tailoring technique, which produces sol-gel processible monomers.

The resulting hybrid glasses show novel optical properties, which would not be expected from those individuals by the loss of individuals' identities after the molecular-level mixing process, thereby creating entirely new properties.

REFERENCES

1. J. Brinker, G. W. Scherer, 1990Sol-Gel Science: The Physics and Chemistry of Sol-Gel Processing: Published by Academic Press, Inc.

2. K. J. Shea, D. A. Loy, O. W. Webster, 1989Chem. Mater., 1, 572.; Shea, K. J., Loy, D. A., & Webster, O. W. (1992) J. Am. Chem. Soc., 114, 6700.

3. R. J. P. Corriu, J. J. E. Moreau, P. Thepot, C. M. M. Wong, 1992Chem. Mater., 4, 1217; Corriu, R. J. P., Thepot, P., & Mang, M. W. C. (1994) J. Mater. Chem., 4, 987.

4. D. A. Loy, K. J. Shea, 1995Chem. Rev., 95 (5), 1431.

5. G. Cerveau, R. J. P. Corriu, C. Lepeytre, 1997J. Orgmetal. Chem., 548 (1), 99.

6. K. M. Choi, K. J. Shea, 1998Photonic Polymer Systems, Fundamentals, Methods, and Application: Edited by D. L. Wise et al. World Scientific Publishing Co. Pte. Ltd.

7. K. J. Shea, D. A. Loy, 2001MRS Bulletin, 26, 368.

8. D. A. Loy, K. . Rahimian, May, 2003Handbook of Organic-Inorganic Hybrid Materials and Nanocomposites: Edited by Hari Singh Nalwa Formerly of Hitachi Research Laboratory, Hitachi Ltd.

9. K. J. Shea, J. Moreau, D. A. Loy, R. J. P. Corriu, B. Bour, 2003Hybrid Materials: Published by Wiley Inter science, New York.

10. L. Zhao, M. Vaupel, D. A. Loy, K. J. Shea, 2008Chem. Mater., 20, 1870.

11. H. W. Oviatt, Jr , K. J. Shea, J. H. Small, 1993Chem. Mater., 5, 943.

12. H. W. Oviatt, K. J. Shea, S. Kalluri, Y. Shi, W. H. Steier, L. R. Dalton, 1995Chem. Mater., 7, 493.

13. L. R. Dalton, 1998Polymers for Electro-Optic Modulator Waveguides in Electrical and Optical Polymer Systems; Fundamentals, Methods, and Applications: Edited by Wise, D. L., Copper, T. M., Gresser, J. D., Trantolo, D. J., & Wnek, G. E., (World Scientific, Singapore) Chapter 18.

14. A. Walcarius, 2001Chem. Mater., 13 (10), 3351.

15. I. Matsubara, 2003AIST Today, 3 (8), 32.

16. D. W. Schaefer, G. Beaucage, D. A. Loy, K. J. Shea, J. S. Lin, 2004Chem. Mater., 16, 1402.

17. Y. X. Pang, S. N. B. Hodgson, J. Koniarek, B. Weglinski, 2006J. of Magnetism and Magnetic Mater., 301(1), March 27, 83.

18. K. M. Choi, K. J. Shea, 1993Chem. Mater., 5, 1067.

19. K. M. Choi, K. J. Shea, 1994J. Phys. Chem., 98, 3207.

20. K. M. Choi, K. J. Shea, 1994J. Am. Chem. Soc., 116, 9052.

21. K. M. Choi, J. C. Hemminger, K. J. Shea, 1995J. Phys. Chem., 99, 4720.

22. K. M. Choi, K. J. Shea, 1995J. of Sol-Gel Sci. Tech., 5, 143.

23. K. M. Choi, J. A. Rogers, 2003J. Am. Chem. Soc., 125, 4060.

24. K. M. Choi, 2005J. Phys. Chem. B, 109, 21525.

25. K. M. Choi, 2007Mater. Chem. Phys., 103, 176.

26. K. M. Choi, K. J. Shea, 2008J. Phy. Chem. C, 112, 18173.

27. J. P. Carpenter, C. M. Lukehart, S. R. Stock, J. E. Wittig, 1995Chem. Mater., 7, 201.

28. N. L. Pocard, D. C. Alsmeyer, R. L. Mc Creery, T. X. Neenan, M. R. Callstrom, 1992J. Am. Chem. Soc., 114, 769.

29. J. H. Kim, P. H. Holloway, 2005Adv. Mater., 17, 91.

30. H. R. Xia, G. W. Lu, P. Zhao, S. Q. Sun, X. L. Meng, X. F. Cheng, L. J. Qin, L. Zhu, Z. H. Yang, 2005J. Mater. Res., 20, 30.

31. S. Mukhopadhyay, K. P. Ramesh, R. Kannan, J. Ramakrishna, 2004Phys. Rev. B: Cond. Matter Mater. Phys., 70, 224202.

32. G. Dantelle, M. Mortier, D. Vivien, G. Patriarche, 2005J. Mater. Res., 20. 472.

33. S. Xiao, X. Yang, Z. Liu, X. H. Yan, 2004J. Appl. Phys., 96. 1360.

34. M. T. Buscaglia, V. Buscaglia, P. Ghigna, M. Viviani, G. Spinolo, A. Testino, P. Nanni, 2004Phys. Chem. Chem. Phys., 6, 3710.

35. S. Stepanov, E. Hernandez, M. Plata, 2004Opt. Lett., 29, 1327.

36. H. Lin, S. Jing, J. Wu, F. Song, N. Peyghambarian, E. Y. B. Pun, 2003J. Phys. D: Appl. Phys., 26, 812.

37. W. Sohler, H. Suche, 2000Opt. Eng., 66, 127.

38. W. Neuman, D. Pennington, J. Dawson, A. Drobshoff, R. Beach, I. Jovanovic, Z. Liao, S. Payne, C. P. J. Barty, 2005Proc. SPIE Int. Soc. Opt. Eng., 5653, 262.

39. X. Huang, Y. Liu, Z. Sui, M. Li, H. Lin, J. Wang, D. Zhao, F. Wang, J. Chen, 2005Proc. SPIE Int. Soc. Opt. Eng., 5623, 679.

40. C. Sanchez, B. Lebeau, F. Chaput, J. P. Boilot, 2003Adv. Mater., 15. 1969.

41. K. J. Shea, D. A. Loy, 2001Chem. Mater., 13, 3306.

42. H. Schmidt, 1989Sol-Gel Science and Technology: Published by World Scientific, Singapore.

43. P. Urquhart, 1988IEEE Proc.-J.; Optoelectronics, 135, 385.

44. C. B. Murray, C. R. Kagan, M. G. Bawendi, 1995Science, 270, 1335.

45. S. A. Empedocles, D. J. Norris, M. G. Bawendi, 1996Phys. Rev. Lett., 77, 3873.

46. L. Armelao, S. Gross, G. Obetti, E. Tondello, 2004Surf. Coat. Technol., 109, 218.

47. D. J. Di Giovanni, 1992Mater. Res. Soc. Proc., 244, 135.

48. R. Rossetti, R. Hull, J. M. Gibson, L. E. Brus, 1985J. Chem. Phys., 82 (l), 552; Zhang, Y., Raman, N., Bailey, J. K., Brinker, C. J., & Crooks, R. M. (1992) J. Phys. Chem., 96, 9098.

49. G. Schmid, 1992Chem. Rev., 92, 1709.

50. J. W. Grate, S. N. Kaganove, S. J. Patrash, R. Craig, M. Bliss, 1997Chem. Mater., 9 (5), 1201.

51. L. Ponson, N. Boechler, Y. M. Lai, M. A. Porter, P. G. Kevrekidis, C. Daraio, 2010Physical Review E, 82, 021301.

52. A. Spadoni, C. Daraio, 2010Proc. Natl. Acad. Sci. USA, 107, 7230.

53. M. A. Porter, C. Daraio, E. B. Herbold, I. Szelengowicz, P. G. Kevrekidis, 2008Physical Review E, 77, 015601(R).

54. P. A. Sullivan, B. C. Olbricht, L. A. Dalton, 2008Journal of Lightwave Technology, 26 (15), 2345.

55. H. , B. Chen, S. Takafumi, L. R. Dalton, A. K. Jen, Y. , 2001J. Am. Chem. Soc., 123, 986.

56. η6 -Bisarene chromium complexes decompose in dilute acid.Oxidation and disproportion products (of both organic and Cr) are observed. [57] η6-Arene Cr(CO)3 complexes form adducts with Lewis and protonic acids but their decomposition products have not been investigated. [58]

57. B. G. Gribov, D. D. Mozzhukhin, B. I. Kozyrkin, A. S. Strizhkova, 1972J. Gen. Chem. USSR (Eng. Transl.), 42, 2521.

58. T. C. Flood, E. Rosenberg, A. Sarhangi, 1997J. Am. Chem. Soc., 99, 4334; Lillya, C. P. & Sahatjian, R. A. (1972) Inorg. Chem., 11, 889.

2

A REVIEW OF COMMUNICATION-ORIENTED OPTICAL WIRELESS SYSTEMS

Deva K Borah, Anthony C Boucouvalas, Christopher C Davis, Steve Hranilovic and Konstantinos Yiannopoulos

1 Klipsch School of Electrical and Computer Engineering, New Mexico State University, Las Cruces, NM, USA

2 Department of Telecommunications Science and Technology, University of Peloponnese, Tripoli, Greece

3 Department of Electrical and Computer Engineering, University of Maryland, College Park, MD, USA

4 Department of Electrical and Computer Engineering, McMaster University, Hamilton, ON, Canada

5 Department of Telecommunications Science and Technology, University of Peloponnese, Tripoli, Greece

ABSTRACT

This article presents an overview of optical wireless (OW) communication systems that operate both in the short- (personal and indoor systems) and the long-range (outdoor and hybrid) regimes. Each of these areas is discussed in terms of (a) key requirements, (b) their application framework, (c) major impairments and applicable mitigation techniques, and (d) current and/or future trends. Personal communication systems are discussed within the context of point-to-point ultra-high speed data transfer. The most relevant application framework and related standards are presented, including the next generation Giga-IR standard that extends personal communication speeds to over 1 Gb/s. As far as indoor systems are concerned, emphasis is given on modeling the dispersive nature of indoor OW channels, on the limitations that dispersion imposes on user mobility and dispersion mitigation techniques. Visible light communication systems, which provide both illumination and communication over visible or hybrid visible/infrared LEDs, are presented as the most important representative of future indoor OW systems. The discussion on outdoor systems focuses on the impact of atmospheric effects on the optical channel and associated mitigation techniques that extend the realizable link lengths and transfer rates. Currently, outdoor OW is commercially available at 10 Gb/s Ethernet speeds for Metro networks and Local-Area-Network interconnections and speeds are expected to increase as faster and more reliable optical components become available. This article concludes with hybrid optical wireless/radio-frequency (OW/RF) systems that employ an additional RF link to improve the overall system reliability. Emphasis is given on cooperation techniques between the reliable RF subsystem and the broadband OW system.

KEYWORDS

optical wireless, personal communication systems, indoor systems, outdoor systems, hybrid OW/RF systems

1. INTRODUCTION

It has been more than 30 years since optical wireless (OW) was proposed as an alternative broadband technology for wireless data transmission applications [1]. The underlying concept of OW is very simple: utilize optical beams to carry data through the atmosphere or vacuum. As a result, OW link architectures are very similar to optical fiber communication point-to-point links, with the exception that no optical fibers are deployed as a transmission medium. OW is also very similar to RF wireless, but radio waves are replaced with light and antennas with free-space optical transceivers. Despite this superficial resemblance between OW and RF links, OW exhibits several appealing attributes when compared to RF. OW links are inherently broadband and optical frequencies in the infrared and visible spectrum are neither regulated nor licensed. Optical components are also cheaper and consume less electrical power than high-speed RF components. Finally, OW links do not suffer from multipath fading and have much less potential for interference with RF-sensitive electronic systems. These advantages do not, however, imply that OW is a universal replacement for RF communications. The application of OW systems is limited when considering area coverage and user mobility, where RF technologies prove invaluable. In addition, OW systems operate under strict eye safety regulations, while at the same time incoherent OW receivers present lower sensitivity than their RF counterparts because of their photo-electric conversion mechanisms and the impact of ambient light noise sources. Table 1 summarizes the main differences between OW and RF systems.

Table 1 Comparison between RF and OW systems

	OW	**RF**
Bandwidth	Not regulated	Licensed
Available line rates	< 10 Gb/s	< 1.25 Gb/s
Path losses	High	High
Multipath fading	No (large collector area)	Yes
Multipath distortion	Only in diffuse indoor systems	Yes
Noise sources	Ambient light	Interference from other users, electrical noise
Detection type	Incoherent	Coherent/Incoherent

	OW	RF
SNR	Depends on optical signal power	Depends on RF signal amplitude
Receiver sensitivity	Low	High
Eye safety	Required	N/A
Electromagnetic compatibility	Yes	Conditional

To better understand the place of OW systems in the wireless world, Figure 1 summarizes state-of-the art commercial RF and OW technologies, as well as technologies under standardization by major bodies including IEEE, 3GPP, Bluetooth and IrDA. Technologies are presented with respect to their area of coverage, ranging from a few centimeters in personal communications to over 1 km in outdoor communications, and the data rates they attain, including low rate legacy links under 1 Mb/s (Bluetooth and older IrDA systems). Clearly, contemporary OW links provide channel rates up to 10 Gb/s, which directly compare to the ones of optical fibers. At the same time, commercial OW links operate at link distances that are challenging to attain in RF (3G/4G) and millimeter-wave (60 GHz) broadband communications.

OW is a unique technology that provides an attractive alternative in niche application areas, complementing fiber-optic and RF wireless solutions when they are either too costly to deploy, create undesirable interference, or are not feasible at all. Figure 2illustrates some of the application areas in which OW has been successfully applied. Two mainstream application areas of OW are last-mile broadband access and office interconnection; both are the business objectives of a number of component and system manufacturers. In such applications, state-of-the-art OW systems support 10 Gb/s Ethernet, which equals the bandwidth provided by metro fiber optic systems and is significantly higher than the 1.25 Gb/s Ethernet provided by competing RF wireless systems that operate in the 60 GHz frequency range. At the same time the deployment cost of OW systems is significantly lower than that of fiber optics, which can easily reach $1M/mile in urban areas. OW installation only requires the alignment of two free-space optical transceivers rather than digging trenches and repairing roads.

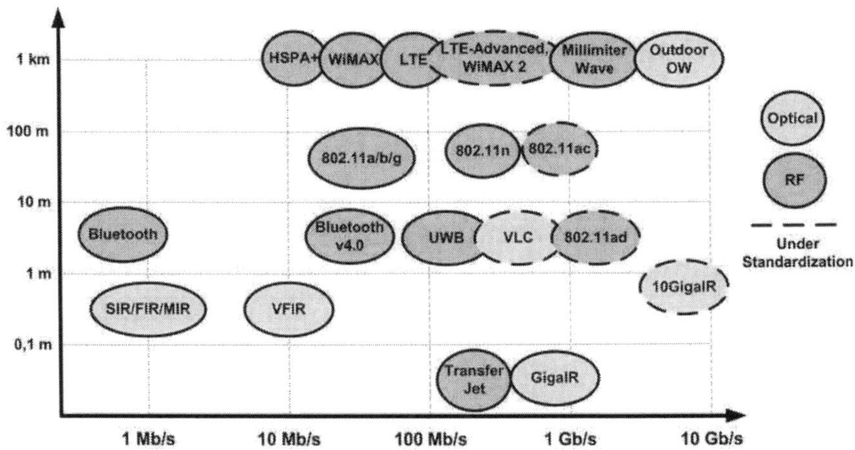

Figure 1 RF and OW technologies as defined in standards and deployed in commercial products.

Another major application area of OW is in personal communication systems. The rapid progress in optical technology over the past 30 years, driven mainly by fiber-optic systems and display technologies, has enabled the mass production of high volume, low cost, and fast optical components that are suitable for short range OW. During the 1990s OW emerged as a candidate technology for data transmission in personal communications systems using protocols developed by the Infrared Data Association [2]. The result of this study was the standardization and commercialization of OW ports that have been extremely popular since the late 1990s and may be found on all kinds of mobile phones and portable computers. The current state-of-the-art in personal communications is Gigabit Infrared (Giga-IR) that operates at link speeds of 512 Mb/s and 1.024 Gb/s. This short-range interconnection is among the highest speed wireless interconnection media available and is easily integrated into portable and consumer devices. At the Expo Comm Wireless Japan in 2008 and 2009, KDDI Corporation demonstrated Giga-IR, which is envisioned as an interconnection medium for future cellular phones.

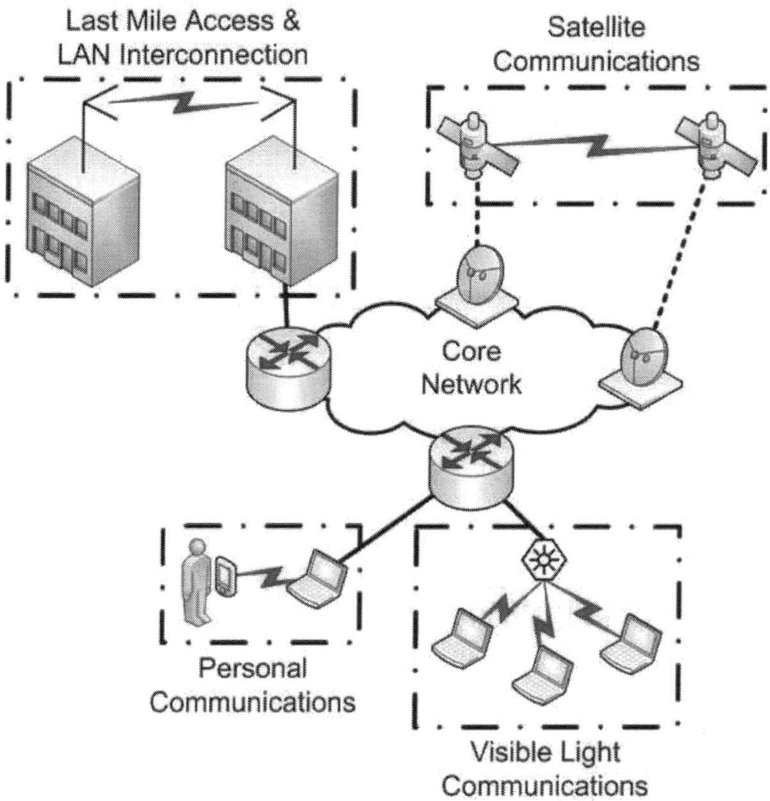

Figure 2 Application areas of optical wireless systems.

Figure 2 also illustrates two additional application areas of OW: indoor and ultra-long-haul systems. OW was initially proposed as a broadband transmission technology for short range diffuse indoor wireless systems [1]. Indoor OW systems are currently being revisited within the framework of Visible Light Communication (VLC) Systems [3] that aim to combine lighting and communications over commercially available visible light LEDs. With respect to ultra-long-haul systems, major technological advances in the fields of adaptive optics and beam acquisition/tracking, which are functions of critical importance in ultra-long-haul outdoor OW links, have enabled the deployment of tractable broadband outdoor OW communication systems. At the time of writing of this article, numerous ultra-long-haul outdoor OW systems have been proposed and demonstrated in a wide range of applications, including terrestrial, earth-to-satellite, earth-to-high altitude platform, intersatellite, and interplanetary communication links [4].

OW is a standalone technology, differentiated from fiber optics and RF wireless as far as usage scenarios are concerned. OW fits perfectly well into the broad scenario of telecommunication networks as a complementary technology that provides communication solutions tailored to diverse application areas. This article aims to present progress in OW systems, underlying principles that lay the basis of OW, and some of the newest trends that appear to be of practical importance in the future. To facilitate the presentation, OW systems have been categorized in four distinct areas: personal communication systems, indoor communication systems, outdoor communication systems, and hybrid OW/RF systems. Personal communication systems are presented in Section 2. The discussion focuses on the most relevant user requirements and standards (IrDA and IEEE), and also includes the Giga-IR extensions that are the latest trend in short range OW systems. Section 3 presents indoor communication systems, with emphasis given to suitable channel models for the diffuse OW link and techniques for achieving spatial coverage. Section 3 also details the latest trend in indoor OW, Visible-Light Communication Systems. Outdoor OW systems are discussed in Section 4. Emphasis is given to the negative impact that the atmosphere has on optical beams, as well as suitable mitigation techniques that largely alleviate this impact. The latest advances in the most challenging application area of outdoor OW systems, satellite, and high-altitude-platform communications, are also presented. This article concludes with Section 5, which presents outdoor hybrid OW/RF systems. The discussion revolves around the cooperation schemes between the OW and RF links that achieve increased system reliability and capacity availability.

2. PERSONAL COMMUNICATION SYSTEMS

Short-range OW has attracted considerable attention for personal communication systems over the past decade, owing to the fact that OW systems offer a viable, low cost and complexity, high bandwidth solution to user-terminal connectivity. OW provides considerable unlicensed (free) bandwidth in the infrared spectrum, the use of which results in significant cost savings to equipment manufacturers since they do not have to pay an extra fee for using this spectrum and their transceiver designs do not have to strictly conform to spectral masks. Additionally, OW is preferable when health, safety, and aesthetics issues are raised to challenge personal communication systems, since its low transmission powers, short link distances, low profile transceivers and limited spatial coverage make OW one of the least obtrusive personal network technologies. Finally, OW neither suffers from nor directly creates RF interference and OW systems can operate side-by-side with RF systems in personal communication networks. The high-speed electronics that drives OW transmissions must be designed not to be a indirect source of RF interference.

Apart from the generic requirements mentioned above, OW has been quite successful in personal communication systems since it exhibits attributes that closely match key requirements in these systems and their applications [2]. From an application perspective, applications like contact information exchange or file transfer between mobile devices are required to have a very short interaction time. In addition, several applications, for instance secure payment, require increased security and immunity to interference. The short distances involved and the limited spatial coverage of OW, which results from the narrow beams that are utilized, provide inherent security, and interference rejection, while at the same time allow for faster completion of the device discovery process that affects the application interaction time. Moreover, the abundant bandwidth of OW transceivers, more than 1 Gb/s at the time of writing of this article, facilitates speedy file transmissions and as a net result the total interaction time is minimized.

The suitability of OW for personal communication systems has drawn the attention of two key players in personal communication systems: IEEE and the Infrared Data Association (IrDA). IEEE had included an OW option in its original 802.11 MAC and PHY standards [5]. The standards defined a wireless communication scheme using diffuse optical channels at 1 and 2 Mb/s. However, they were never updated and are now rendered obsolete. Recently, IEEE launched a separate Task-Group within the framework of 802.15 (IEEE 802.15.7), which is responsible for producing MAC and PHY standards for OW communications based on visible light [6]. IEEE activities will be further discussed in the next section that details indoor systems.

The IrDA aims at equipping personal communication devices (computers, cellular phones, PDAs, etc.) with low-cost and high-bandwidth OW equipment. To this end, IrDA has produced numerous standards for the definition and implementation of (a) an infrared physical layer, (b) a protocol stack that is well suited for short-range OW, and (c) several application scenarios. Even though IrDA had produced standards for multiple-access implementations over infrared links (Advanced Infrared--AIr) [7–9], with possible applications in personal and local area networks (PANs and LANs), it soon became evident that it will be quite challenging for OW to compete with RF in terms of area coverage and user mobility. As a result, IrDA has focused primarily on point-and-shoot usage scenarios, since OW is well suited for this [10]: point-and-shoot applications require ultra-fast connection establishment and data transfer; both are readily offered by OW. The rest of the section focuses on the IrDA protocol stack, compliant applications and a comprehensive description of related research efforts.

2.1. The IrDA protocol stack and related applications

The IrDA protocol stack is shown in Figure 3. The lower layer is the physical layer [11, 12] (IrPHY), which operates at a nominal wavelength of 880 nm, although in practice infrared transceivers are allowed to operate in a range that extends from 850 to 900 nm. LEDs and silicon PIN photodiodes are utilized to reduce the system cost to an absolute minimum at lower speeds, while lasers are a more costly alternative at higher speeds. The average optical powers transmitted by lasers and LEDs are limited by eye and skin safety standards, according to [13]. Transmission (line) speeds vary from a couple of kb/s to 1 Gb/s and IrDA has provided detailed physical layer specifications for the following communication speeds:

- Serial Infrared (SIR): SIR is 16550 UART compatible, providing link speeds at 2.4, 9.6, 19.2, 38.4, 57.6, and 115.2 kb/s. The bytes that are to be transmitted over the infrared link are first placed in a 16550 UART chip, which is used to modulate the driving current at the transmitter LED. The UART chip appends a start and a stop bit to the transmitted byte, inverts the bit pattern and also alters the duty cycle of the '1's in the transmitted 10-bit UART frame to 3/16.

- Medium Infrared (MIR): MIR speeds are 0.576 and 1.152 Mb/s. Line coding is almost identical to the SIR speeds, with the only exception being the duty cycle of the '1's in MIR speeds (1/4 instead of 3/16).

- Fast Infrared (FIR): FIR supports a single line speed of 4 Mb/s. Line coding in FIR speeds follows a pulse-position-modulation (PPM) scheme with two-bit symbols. Four chips are required to transmit all possible symbols, and the symbol to chip mapping is defined as {00,1000}, {01,0100}, {10,0010}, and {11,0001}.

- Very Fast Infrared (VFIR): VFIR supports 16 Mb/s only. Line coding in VFIR follows a run-length limited HHH(1,13) scheme, which guarantees at least one empty chip after each pulse and allows for a maximum of 13 empty chips between pulses. Under the HHH(1,13) coding scheme, data bit pairs are encoded into three-chip symbols using the state transition table detailed in [11].

- Gigabit Infrared (Giga-IR): Giga-IR supports 512 Mb/s and 1.024 Gb/s. Line coding in Giga-IR is based on the 8010B DC-balanced encoding scheme, in which a data byte is encoded on a 10-bit character, so as to facilitate clock recovery at the receiver and to meet the stringer timing requirements of Giga-IR links. The Giga-IR optical transmitter can be either a LED or LD: if a LED is used as the transmitter then the encoded characters are sent over the infrared link using a four-level ASK modulation format, otherwise if a LD is used as the transmitter then the modulation format is a two-level ASK [12].

IrDA has also provided recommendations on a number of infrared link parameters. In SIR to VFIR speeds the maximum distance between devices should not exceed 1 m (or 200 mm for low power transceivers), while the 3 dB half-width angle ranges between 15 and 30° at the transmitter and 0-15° at the receiver. In Giga-IR speeds a docking station is typically used and the maximum distance is limited to 60 mm, while the transceiver angles are also limited to 5°. Moreover, the bit-error-rate that should be achieved for all infrared links is no more than 10^{-8} for SIR to VFIR speeds and no more than 10^{-10} for Giga-IR, comparing directly to the BERs achieved in modern optical fiber networks. Additional optical and electrical parameters including (a) maximum/minimum optical intensity at the transmitter and optical irradiance at the receiver, (b) pulse rise/fall time, (c) pulse jitter, (d) minimum/maximum pulse duration, and (e) maximum allowable transmission rate deviations may be found in [11, 12] for all speeds ranging from SIR to Giga-IR.

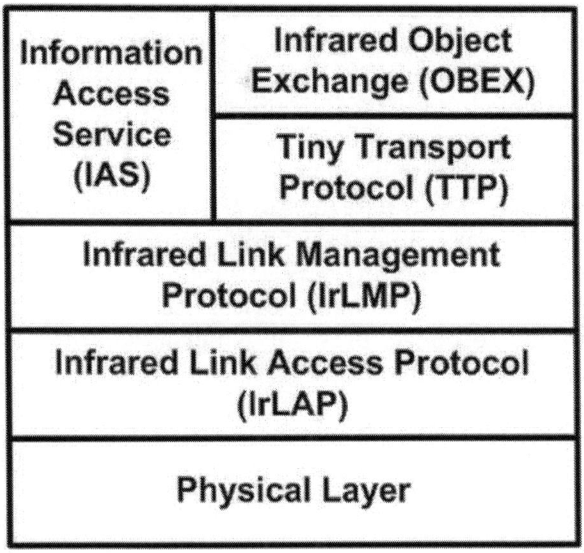

Figure 3 The IrDA protocol stack.

The IrDA protocol stack operates above IrPHY and consists of four main transfer protocols: the Infrared Link Access Protocol (IrLAP) [14], the Infrared Link Management Protocol (IrLMP) [15], the Tiny Transport Protocol (TTP) [16], and the Object Exchange (OBEX) protocol [17]. IrLAP is an HDLC-based link layer protocol with a primary function to provide reliable data transfer over half-duplex OW links. Earlier IrLAP implementations (SIR to VFIR speeds) operated in a Go-back-N fashion and maintained frame sequencing information in both directions. Currently, IrLAP is being modified to take full advantage of the bandwidth provided by Giga-IR speeds [18], and the next IrLAP standard is expected to provide (a) extended frame size, (b) extended window size, and (c) selective-repeat mode of operation. Apart from reliable data transfer at Gb/s speeds, IrLAP has additional features to account for the volatile nature of the infrared physical layer, including extended and dynamic node addressing, link recovery and rapid node discovery (25-300 ms) if fast-connect options have been implemented in IrLAP [19].

IrLMP operates above IrLAP and is responsible for providing upper layers with a multiplexing scheme over IrLAP, as well as a link control mechanism and a service discovery mechanism. IrLMP in fact comprises two separate entities, the Link Management Multiplexer (LM-MUX), which is responsible for link control and data multiplexing and (b) the Link Management Information Access Service (IAS) that maintains information about the services that have been registered at the local station. IrLMP communicates with upper layers and the IAS through multiple Logical Service Access Points (LSAPs). Each LSAP is assigned with a unique identifier that is called the Logical Service Access Point Selector (LSAP-Sel), and source/destination LSAP-Sel pairs are used to multiplex/demultiplex traffic from multiple IrLMP connections in LM-MUX. Moreover, LSAP-Sels of local services are registered in IAS to facilitate service discovery. Remote devices access the IAS by means of a simple command-response protocol (Information Access Protocol--IAP); the LSAP-Sel of IAS itself is well known and equals 0 × 00.

TTP is heavily associated with IrLMP and its main functions are (a) to provide per IrLMP connection flow control, and (b) segmentation and re-assembly (SAR) features. Even though IrLAP provides lower level flow control (back-pressure), it is not capable of distinguishing between individual IrLMP flows and as a result using IrLAP back-pressure will slow down all IrLMP connections. TTP on the other hand implements per-connection flow control by associating its own Service Access Points (TTPSAPs) with the LSAPs that are present in the LM-MUX on a one-to-one basis. The flow control mechanism itself is credit based: the TTP endpoints issue credits (up to 127 initially) to each other and they use up a credit whenever they send a TTP frame or receive a new credit whenever a TTP frame has been received at the other end. If a TTP endpoint reaches zero credits, it must wait for additional credits to be issued from the other side before sending additional frames.

OBEX is the session protocol that has been accepted by IrDA to ensure interoperability between devices, and provide a communication API for user applications. OBEX comprises three main components: (a) a model for representing objects, as well as information about objects, (b) a session protocol that enables session layer communication between devices, and (c) an application framework. Similar to HTTP, OBEX uses a header-based model to describe both the object attributes and the actual data of the object itself. Session peers, one acting as a client and the other as a server, exchange header information using a request/response communication protocol. Typically, OBEX operates in a stop-and-wait fashion, which means that only a single OBEX frame may be outstanding. This behavior limits the infrared link utilization, since requests and responses that involve the transmission of multiple OBEX frames need to wait for an equal number of link turnarounds. In order to mitigate this effect, enhanced OBEX implementations allow for a Single Response Mode (SRM), in which multi-frame requests/

responses may be acknowledged by a single frame. Finally, the primary element of the application framework is the default OBEX server. This is the "well-known OBEX server" that is registered in the IAS database and provides basic functionalities like accessing file systems, databases, services and applications. The default OBEX server uses an Inbox service, so that clients can exchange objects using a standard location. Alternatively, clients may also communicate with non-default OBEX servers; in such case multiple OBEX server instances are running along with the applications. The non-default server instances are registered with different LSAP-Sels in the IAS, so that there is no confusion with the default OBEX server.

IrDA has also standardized a stripped-down version of the protocol stack to facilitate implementations in limited capability terminals. This version is commonly known as IrDA Lite and maintains only the most fundamental operations of the protocols involved [20]. A more recent variation of the protocol stack, IrSimple [21], achieves faster transfer speeds by allowing blocks of information to be exchanged over IrLMP. Since the TTP flow control mechanism does not allow for block sizes over 127 frames, TTP is replaced by IrSMP (Sequence Management Protocol for IrSimple) in the IrSimple protocol stack [22]. IrSimple maintains the SAR capabilities of TTP and implements a frame sequencing mechanism to ensure the successful transmission of frame blocks.

IrDA has also provided standards for application related frameworks. One such application framework is Infrared Financial Messaging (IrFM), which supports digital payment applications running on infrared-enabled personal communication devices [23]. In IrFM, a personal communication device (also called a Personal Trusted Device--PTD) interacts with a Point of Sale (POS) terminal. Applications running on PTD and POS communicate over OBEX, with PTD being the OBEX server and POS being the OBEX client. Applications may not access OBEX directly, however, but only through IrFM services. To this end, the IrFM standard defines a number of core services that enable basic transactions between the POS and the PTD, as well as a Core IrFM Protocol (CIP) that passes transaction commands from IrFM services to OBEX. A key feature of IrFM is its fast transaction mechanism (point-and-pay), as a direct result of the point-and-shoot nature of the IrDA protocol stack. In addition, IrFM has been developed to maintain simplicity, security, interoperability and reliability in infrared financial transactions.

Another application framework is "wire replacement" and refers to the replacement of parallel and serial cables with IrDA links. So far there have been two relevant standards, the Serial and Parallel Port Emulation over IR (IrCOMM) [24] and the Infrared Universal Bus (Ir-USB) [25]. IrCOMM defines the emulation of parallel and serial ports over the IrDA protocol stack. The drive between the emulation is to facilitate the migration of applications that have been initially written to work over COM and LPT ports to IrDA compliant devices. To this end, IrCOMM differentiates between four service types: three-wire-raw, three-wire, nine-wire and Centronics. Three-wire-raw involves no control information and the role of IrCOMM is to fill up payload of IrLMP frames with user data. The other three service types, however, require a separate control channel and as a result IrCOMM frames carry control information along with the user data. In addition, IrCOMM interprets flow control commands and utilizes the credit based flow control mechanism of TinyTP in an appropriate fashion. Ir-USB, on the other hand, is mainly based on the USB 2.0 protocol stack and utilizes only IrLAP and IrPHY (Giga-IR in particular) as an ultrafast transmission mechanism. As a result, an operational speed of 480 Mb/s is provided to USB clients over a 512 Mb/s or 1.024 Gb/s infrared link. The critical subsystem of the Ir-USB architecture is the Ir-USB controller that acts as an interface between the electronic USB controller and IrLAP. Apart from transmitting data at very high speeds, the Ir-USB controller is also responsible for handling special purpose USB packets, so that the infrared link is essentially transparent to the electronic USB controller.

2.2. Research-related literature

OW personal communication systems involve point-to-point links with distances that are typically limited to a few tens of millimeters and very narrow optical beams. As a result multipath dispersion and fading, which severely affect the performance of indoor and outdoor systems, respectively, are not present in personal communication systems. The dominant impairment in personal communication systems is optical noise and interference from the sun and artificial lighting [26, 27], still the narrow reception angles filter out most of this. The reader may refer to Section III.D for a more detailed discussion on noise and interference, as well applicable mitigation techniques.

The optical channel in personal communication systems is considered to be linear, time invariant and non-dispersive [1, 28]. In practice, there are two main goals regarding the physical layer in personal communications systems: (a) to attain increasingly high link speeds, and (b) to deploy suitable codes that facilitate receiver functions such as clock and frame recovery. To these ends, simple amplitude modulation formats like OOK and amplitude shift keying (ASK) have been previously presented and evaluated [27], even though more complex phase and frequency modulation formats are also possible [29]. Relevant modulation schemes include PPM [30], differential PPM (DPPM) [31], digital pulse interval modulation (DPIM) [32], run-length limited (RLL) coding [33] and DC-balanced coding [34]. PPM is advantageous over OOK for low rate systems, since it requires decreased average power and is more robust against optical noise, especially noise components near baseband resulting from fluorescent lamps [27]. Still, PPM requires a bandwidth increased by a factor of $M/\log_2 M$ as compared to OOK, where M is the number of chips that are required to construct the PPM symbol. DPIM and DPPM require less bandwidth (approximately half) than PPM and more average power [30], but they also occupy more bandwidth than OOK and ASK. ASK is better suited for higher rate systems; it utilizes the same bandwidth as OOK to provide a $\log_2 M$ times faster line rate, at the expense of increased average optical power. Table 2 and Figure 4 summarize the bandwidth and BER performance of PPM variants and ASK systems in short-range OW.

RLL codes, on the other hand, are used to alleviate timing uncertainties that cause erroneous data recovery in high-speed PPM receivers. Specifically, the VFIR IrDA HHH(1,13) run-length limited code performs better clock and data recovery than PPM, while maintaining an acceptable bandwidth efficiency (approximately 67%) [33]. DC-balanced coding is also receiver oriented and is used to mitigate the baseline wander effect caused by high-pass filtering of optical noise at the receiver [34–36]. More sophisticated modulation schemes that involve transmission over multiple carriers, such as subcarrier modulation (SCM) [37, 38] and orthogonal frequency division multiplexing (OFDM) [39–41] are also possible. However, SCM and OFDM are primarily designed to minimize inter-symbol (ISI) interference in indoor systems, and therefore require more complex coding and modulation hardware at the transceivers when compared to OOK.

Table 2 Bandwidth and BER performance of OOK and PPM variants

Modulation format	Data rate	BER
ASK	$R_b = B \cdot \log_2 M$	$2 \cdot \frac{M-1}{M \cdot \log_2 M} \cdot Q\left(\frac{1}{M-1} \cdot \sqrt{SNR \cdot \log_2 M}\right)$
PPM	$R_b = B \cdot \frac{\log_2 M}{M}$	$\frac{M}{2} \cdot Q\left(\sqrt{SNR \cdot \frac{M \cdot \log_2 M}{2}}\right)$
DPPM	$\overline{R_b} = 2 \cdot B \cdot \frac{\log_2 M}{M+1}$	Variable symbol duration, refer to [31] for a detailed presentation
DPIM	$\overline{R_b} = 2 \cdot B \cdot \frac{\log_2 M}{M+3}$	Variable symbol duration, refer to [32] for a detailed presentation

Bandwidth B is the first-null bandwidth, SNR is the electrical signal-to-noise ratio and M is the number of chips that construct a symbol in PPM or the number of power levels in an ASK system. Function $Q(x)$ is defined as $Q(x) = \frac{1}{\sqrt{2\pi}} \cdot \int_x^{\infty} e^{-u^2/2} \cdot du.$

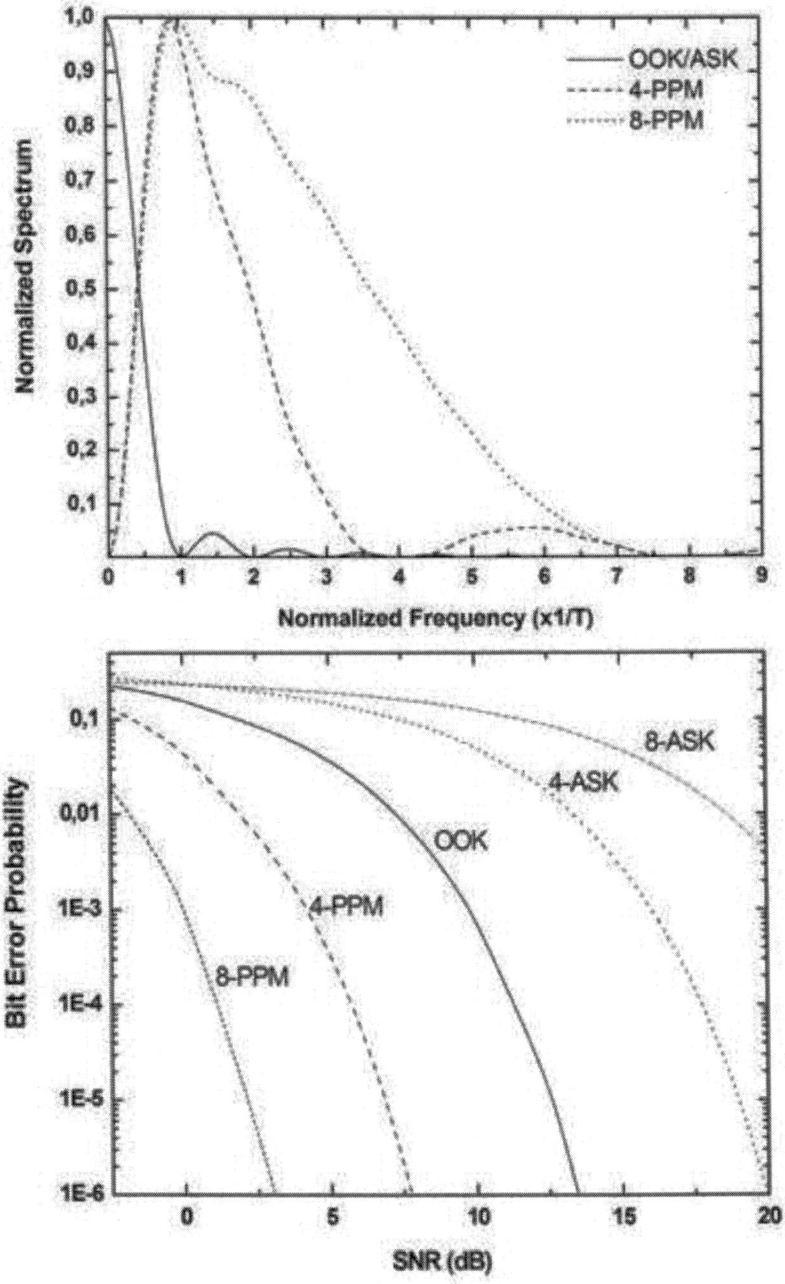

Figure 4 Spectral properties and BER performance of PPM and OOK/ASK systems, following the analysis of [60].

With respect to protocol analysis in OW personal communication systems, both the initial IEEE 802.11 and IrDA AIr standards have been previously studied in the literature. A performance analysis for AIr may be found in [42–44] where the link-layer frame size is contrasted against the average frame delay and system throughput. IEEE 802.11 has been extensively investigated in the literature, but a complete overview is beyond the scope of this article. Two articles that evaluate the performance of initial 802.11, and apply to OW MAC and PHY are [45, 46]. As far as the point-and-shoot oriented IrDA protocol stack is concerned, numerous research efforts have been presented in the literature. A primary research topic involves the determination of the optimum values that maximize the system throughput for network parameters including (a) the OBEX frame size, (b) the TTP window size, and (c) the IrLAP window size, frame size and link turnaround time. The optimum IrLAP frame and window sizes have been calculated in [47, 48] as a function of the link BER, speed and turnaround time. Cross-layer optimization that involves the TTP, however, shows that the IrLAP network parameters are not independent of the TTP flow-control mechanism [49], and this owes to the fact that TTP frame streams are not temporally continuous. Moreover, the Stop-and-Wait fashion in which OBEX operates further complicates the determination of optimum values for the OBEX, TTP and IrLAP network parameters [50, 51]. Further analysis of the IrDA protocol stack involves several variations: the suitability and scalability of IrSimple for transmission of relatively small files over high-speed OW links is discussed in [52]. An alternative session protocol that is suitable for the transmission of larger files, IrBurst [53], is also optimized against the TTP and IrLAP network parameters in [54, 55]. Finally, a detailed analysis of IrFM attributes, including its connection delay (transaction speed), simplicity, interoperability, security and reliability, along with a comparison between IrFM and competing technologies may be found in [56].

3. INDOOR OW SYSTEMS

OW has been extensively studied as a broadband, low cost and power efficient technology for indoor communication systems [1,27, 57–62]. Similar to the case of personal communications, OW indoor systems provide considerable cost savings to device manufacturers and end users due to the existence of unlicensed available spectrum, are perceived as safer than competing RF technologies to humans and do not suffer from RF interference. Indoor OW link architectures, however, are quite different from their personal communication counterparts due to the fact that a key requirement here is user mobility. In real-world indoor systems the user is allowed to move within a limited area of coverage, which typically equals the size of a room or an aircraft cabin.

Providing OW access to a moving user can be challenging, since the optical beams are blocked by objects inside the room. A possible solution is to deploy a roof-mounted and narrow-beam transmitter that is controlled by a tracking mechanism. The tracking mechanism rotates the transmitter and directs the narrow optical beam from the transmitter to the mobile receiver. This approach leads to a point-to-point OW link whose architecture is very similar to that of personal communication systems (Figure5a). An alternative approach is to design the transmitter with a broad-angle emission pattern and allow multiple reflections from the walls and objects inside the room. In such a case, the OW system is categorized as diffuse (Figure 5b). If the emission pattern is properly engineered, the multiple reflections result in uniform illumination inside the area boundaries, even at places that are out of line-of-sight of the transmitter, and therefore enhance user mobility [63]. However, the multipath propagation in diffuse channels leads to time-dispersion of any pulse transmitted on the channel. Multipath dispersion is a critical impairment, since it drastically reduces the system bandwidth, and attainable data rate, and increases the link losses. As a result, point-to-point and diffuse topologies present a tradeoff between mobility and bandwidth. More advanced topologies, for instance multispot diffusing (Figure 5c), combine the best of both worlds and allow for significant user mobility while at the same time achieve two orders of magnitude higher bandwidth than purely diffuse ones. All three topologies (point-to-point, diffuse, and multispot diffusing) are discussed in further detail in Sections 3.1-3.3, and a summary of their respective strengths and weaknesses is given in Table 3.

Another major difference between personal communication and indoor systems is the impact of noise on their performance. Indoor OW receivers are typically designed with broader reception angles and as a result the received optical noise power levels are higher when compared to those of personal communications. Noise originating from the sun and artificial lighting interferes with the optical signal and degrades the signal quality at a baseband frequency range between DC and tens of MHz. This is unfortunately the operational frequency range for most indoor OW systems. Filtering techniques combined with modulation formats are required to reject a significant percentage of the received noise power and to limit the noise induced performance degradation. Noise sources, their impact on indoor OW and mitigation techniques are discussed in Section 3.4.

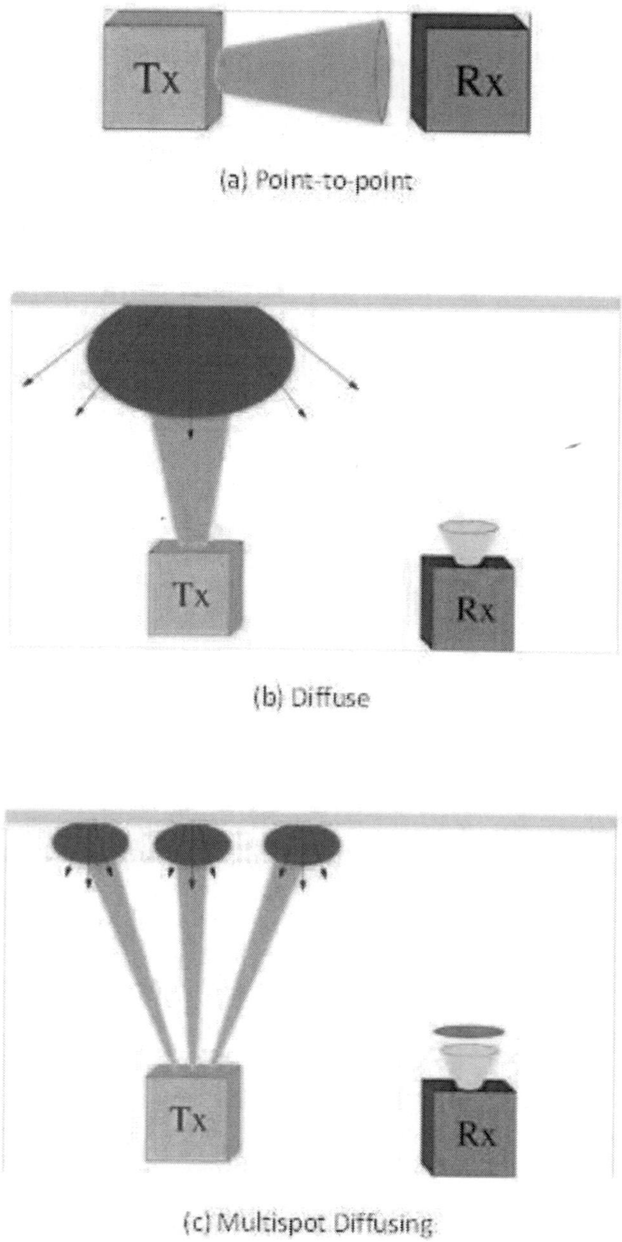

Figure 5 Topologies for indoor optical wireless systems. (a) Point-to-point, **(b)** diffuse, **(c)** multispot diffusing.

Table 3 Comparison of OW link topologies for indoor systems

	Link type		
	Point-to-point	**Diffuse**	**Multispot diffusing**
Link rate	High	Moderate	Moderate
Implementation complexity	Low	Low	High
Beam pointing	Required (exact)	Not-required	Required (partial)
Beam blocking	Yes	No	No
User mobility	Limited	High	High
Dispersion	None	High	Low
Path loss	Low	High	Moderate
Impact of ambient noise	Low	High	Low

Finally, apart from discussing system requirements and key impairments in indoor OW systems, this section also presents the latest trend in indoor OW: VLCs. The main idea behind VLCs is to use solid state LEDs, which were originally developed for indoor lighting, for communication purposes, as well. VLCs are particularly appealing since they do not require additional transmitters and receivers, as is the case in infrared communications, and since ceiling LED installations are already designed to provide a uniform illumination pattern of high optical power within the area of coverage [64]. VLCs are presented in Section 3.5.

3.1. Point-to-point links

Point-to-point links require a line-of-sight (LOS) between the transmitter and the receiver, as shown in Figure 5a. Such a link topology permits the use of highly directional transmitters with low beam divergence and narrow field-of-view (FOV) receivers. Thus, point-to-point links are able to reject the majority of ambient light that interferes with communications. In addition, the directionality of the link removes the impact of multipath dispersion since any reflected transmitter energy is essentially rejected by the receiver. Still, a disadvantage of this link topology is the fact that precise alignment and pointing are required, making the realization of mobile links difficult.

Very high data rates can be achieved in point-to-point indoor OW links by means of spatial division multiplexing, a technique that has been considered for some time. In [57], hundreds of parallel links, each operating at 50 Mb/s, were shown feasible in large areas using a ceiling-mounted transmitter. The transmitter was capable of creating a large number of narrow beams pointed to a variety of locations. Although such systems are ideal for high-speed data distribution, they are of high complexity and do not permit motion of terminals. High-speed electronic tracking was reported in [65] to support mobility at a rate of 155 Mb/s over ranges of nearly 2 m. Tracking was accomplished by using a transmitter that had an array of laser diodes combined with optics and a receiver with a wide FOV and detection array. Tracking involved the selection and switching of paths onto the appropriate receive element.

Point-to-point OW links have also found a great deal of popularity in a variety of shorter-range applications. As an example, the IrDA IrFM specification permits secure and quick payment options at point-of-sale terminals [56], as has been detailed in Section 2. Another example is a system provided by Talking Signs, which is able to deliver voice to visually impaired individuals using fixed infrared transmitters and hand-held receivers. Yet another area of application that has been active for some time is the use of OW links for high-speed backplanes. High bandwidth communication between circuit boards is a major limitation in many large computing and communication platforms. Using an array of tightly aligned transmitters and receivers, high-data rate links that are free from the capacitive, EMI and bandwidth limitations of on-board links [66, 67] can be established. An OW backplane has also been proposed to provide intra-craft data connectivity in a spacecraft [68].

The majority of indoor point-to-point OW links have single element transmitters and receivers. However, great gains can be attained by using multiple-transmitters and receivers in a coordinated fashion. The implementation of such Multiple-Input-Multiple-Output (MIMO) links involves the use of laser and photodiode arrays [69], optical interconnects [70] or a spatial-light modulator (SLM) as a transmitter and an imager as a receiver. Incoming data is either parallelized and transmitted over multiple beams, while when a SLM is employed, data is sent by creating a series of two-dimensional (2D) optical intensity images. OW MIMO links require tight spatial alignment between the transmitter and receiver arrays to avoid interference between the channels [71–73]. Recently a spatial processing technique that does not require tight spatial alignment has been developed for OW MIMO links [74,75]. In this technique, the interference between spatial channels is modeled and used in communication system design directly to provide high spectral efficiency. Alternatively, interference can be minimized by allowing only a single laser element to transmit, as proposed in [76], or by using interference cancelation techniques at the receiver [77].

MIMO links have been used in a variety of applications including 2D barcodes [78, 79], page-oriented recording [80, 81], holographic storage [71, 72, 82, 83], and MIMO OW communications [27, 66, 84, 85]. With respect to the 2D barcodes, OW MIMO links are currently used on a wide array of cellular telephones in the form of barcode readers. Data are encoded as a 2D barcode which is either printed or displayed on a billboard. A cell phone camera is then used to image the barcode and processing is done to extract the data. Many newspapers now publish 2D barcodes which, when imaged by a cell phone camera, direct the cell phone browser to the appropriate webpage for additional details on the story [86].

3.2. Diffuse indoor OW links

Initial studies into diffuse OW links were presented in 1979 [1] and diffuse links were shown to be robust, immune to shadowing and allowed for terminal mobility. Unlike point-to-point OW links, a diffuse OW link does not require the existence of a LOS to establish communications. Rather, as illustrated in Figure 5b, it employs wide beam transmitters and large FOV receivers to exploit diffuse reflections from the walls and objects inside the room and establish a communication link [1, 27, 87, 88]. As a result, there is inherent path diversity and diffuse links are relatively immune to blockage and pointing errors and permit a great degree of mobility for the receivers. However, the received signal is corrupted by multipath dispersion due to the large number of collected reflections at the receiver. The temporal spread of reflections results in inter-symbol interference at high data rates and limits the system bandwidth. Notice that although diffuse OW links suffer from multipath dispersion, they do not exhibit multipath fading as in RF links. The detector in diffuse OW links typically has a size of many thousands of wavelengths and the received optical intensity is integrated over the surface of the detector and provides an inherent degree of spatial diversity [27]. Thus, the channel characteristics are unaltered when moving the detector by many wavelengths.

The indoor diffuse OW channel is well modeled as a linear, time-dispersive channel with a time-invariant impulse response $h(t)$ [89]. The impulse response is calculated from the individual reflections of walls and objects inside the room and there is an array of methods in the literature to compute the impulse response in optical intensity domain for a typical room layout. These methods include ray tracing [89, 90], experimental measurements [28, 91, 92], simplified functional modeling [93, 94], photon tracing [95] and Monte Carlo ray tracing methods [96–99]. While specular reflections are possible, the vast majority of reflections are diffuse since the dimensions of the surface roughness is on the order of the wavelength of light used in the link [95]. As a result, the optical power is reflected nearly independently of the angle of incidence of the optical wave. The intensity distribution of an individual diffuse reflection is well approximated by a Lambertian pattern [1, 60, 89].

The impact of multipath dispersion on diffuse indoor OW links is quantified by the path loss and normalized delay spread. The path loss is calculated from the channel impulse response $h(t)$ and ranges from 10^{-7} to 10^{-5}[27, 28]. The temporal dispersion due to multipath can be measured by the root-mean square (RMS) delay spread of the impulse response and ranges from 2 to 10 ns [27, 28]. A popular and convenient model that correlates the temporal dispersion with the channel bandwidth is the exponential-decay model [93, 94]. In this model, the 3-dB bandwidth of the channel impulse response is calculated from the channel temporal dispersion D using the following simple analytical relation

$$f_{3\text{dB}} = \frac{1}{4\pi D}. \tag{3.1}$$

A range in temporal dispersion D from 2 to 10 ns [27, 28] implies that the 3-dB channel bandwidth approximately ranges from 10 to 40 MHz. The impulse response, delay spread and path loss depend on the particular composition and arrangement of the room and are considered fixed once the room characteristics are fixed [27].

Multipath dispersion severely limits the diffuse channel bandwidth and a vast array of modulation, coding and equalization techniques have been applied to combat this effect [27, 89, 100–103]. An important aspect of these techniques is that amplitude constraints must also be imposed on all signals transmitted on OW channels, since the optical power signal must satisfy a non-negativity constraint. In addition, due to eye- and skin-safety requirements a limitation on the average optical power is also mandatory as imposed by international eye-safety limitations [13]. This power limitation is in stark contrast to the mean squared constraint in RF channels. As a result, the use of modulation and coding for RF channels directly on OW channels is seldom efficient and the unique power and bandwidth efficiency requirement of the OW channel must be considered in any communication system design [60].

Diffuse OW links have been standardized by both IrDA and IEEE. The standardization of diffuse OW links from IrDA resulted in the Advanced Infrared (AIr) physical layer interface that defines the operation of 4 Mb/s diffuse OW links inside typical room sizes [2, 43, 44, 104]. IEEE included a physical layer specification for infrared OW communications at 1 and 2 Mb/s in its 802.11 standard. Experimentally, diffuse infrared OW links have been demonstrated at 25 Mb/s [105] and 50 Mb/s [106, 107].

3.3. Multispot diffusing links

Spatial diversity techniques have been successfully applied to indoor OW links in a topology that is termed multispot diffusing (MSD) and shown in Figure 5c[108]. The MSD architecture aims to combine the advantages of point-to-point links with the mobility of diffuse links [108–115]. To this end, the MSD transmitter modulates data onto a series of beams that are projected onto the ceiling above the communications floor. The MSD receiver ideally images one or perhaps several spots and decodes data from the diffusely reflected energy. The same data is modulated for all spots and the arrangement and number of spots is optimized so that at least one spot is in the imager for every receiver position.

Even though both diffuse OW and MSD links rely on diffuse reflections on surfaces inside the room, there are large differences in the underlying channel models of both topologies. The first main difference lies in the intensity distribution of the optical transmitter. Whereas diffuse OW links emit light over a large divergence angle, the MSD transmitter uses a series of narrow divergence beams directed to the ceiling. This more efficient organization of the optical intensity leads to a far smaller path loss in MSD systems than in diffuse systems. The second major difference lies in the construction of the receiver. The MSD receiver consists of a series of narrow FOV elements directed to the ceiling. These narrow FOV receivers reject a large portion of ambient light. In addition, the narrow FOV receivers are able to reject any stray multipath signals and this effect significantly improves the bandwidth of the channel. It has been shown that in a typical room setting the channel bandwidth of an MSD system is in excess of 2 GHz, which is in stark contrast to the tens of MHz available in diffuse systems [116].

One of the challenges in implementing an MSD link is the creation of an array of spots on the ceiling. Although straightforward, using arrays of emitters with their own optics is bulky and expensive [115]. A less expensive and compact method is to use a laser diode and a computer generated hologram. In [112], a hologram which produces an array of 8 × 8 beams is presented, while holographic diffusers with fewer lobes were considered earlier in [63]. Holograms have also been considered to create arrays as large as 10 × 10 spots [117]. Other work has considered different geometries for MSD spots including a line strip and a diamond in an effort to reduce transmitter complexity [118, 119]. Experimental work on MSD systems has also shown promising results. An angle diversity receiver composed of discrete elements working at 70 Mb/s at a BER less than 10^{-9} over a 4-m range was reported in [115]. In [114], system issues for a 100 Mb/s MSD link were also presented. A 140 Mb/s MSD system with a single receiver [120] and a 155 Mb/s using a multi-element imaging receiver are presented in [65]. Following the above, MSD links maintain the high bandwidth and low path losses of point-to-point links and at the same time achieve mobility that is comparable with those in diffuse links at the expense of slightly more complicated transmitters and receivers [112]. As a result, the MSD architecture is an excellent topology for future indoor high-speed content distribution systems.

3.4. Noise and interference modeling and mitigation

Noise and interference are two additional impairments that affect the performance of indoor OW links. In typical indoor diffuse OW systems, which involve intensity modulation and direct detection (IM/DD), a good model that describes noise and interference is the baseband model

$$y(t) = R \cdot x(t) * h(t) + z(t). \tag{3.2}$$

$x(t)$ is the transmitted optical power, $y(t)$ is the current at the receiver, $h(t)$ is the optical channel impulse response, R is the receiver photodiode responsivity and * denotes continuous time convolution. The signal $z(t)$ represents noise and interference added at the receiver. In general, $z(t)$ is a high intensity shot noise process. Such processes are well modeled as zero-mean, Gaussian distributed, independent of $x(t)$ and white [27]. However, this model is only true under the assumption of intense ambient light. For example, in links with low background illumination a photon-counting model is more appropriate [60].

The sources of noise and interference include background illumination, lighting from incandescent and fluorescent sources as well are front-end circuit noise in the preamp of the receiver [1, 26, 27, 60, 87, 121, 122]. Figure 6 summarizes the spectral distribution of solar, incandescent, and fluorescent radiation in the visible and near-infrared wavelength ranges, which are typically employed in OW systems. The sun emits broadband un-modulated light that is converted to shot-noise current at the receiver. The root-mean-square (RMS) of the noise current is proportional to the incident optical power and the receiver bandwidth and band-limited optical and electrical filters at the receiver can reject a significant portion of the shot noise. Incandescent lamps are modulated by their AC power supply (50/60 Hz) and produce a narrowband photocurrent with a total bandwidth of less than 2 kHz [122]. In practical OW systems the receiver bandwidth is significantly higher than 2 kHz and the photocurrent from incandescent light is modeled as a constant RMS shot noise current that is added to the photocurrent from sunlight [27]. Fluorescent lamps, on the other hand, emit light that is modulated at the power supply frequency (50/60 Hz) or higher frequencies (tens or hundreds of kHz). Light from 50/60 Hz fluorescent lamps produces a photocurrent with a total bandwidth of 20 kHz [122], while higher modulation frequencies result in photocurrents that have a considerable part of their power spectrum in the MHz range [121]. Consequently, the bandwidth of the photocurrent from fluorescent lamps is a significant percentage of the total bandwidth in practical OW systems and the photocurrent is modeled as a time-varying interference signal [27].

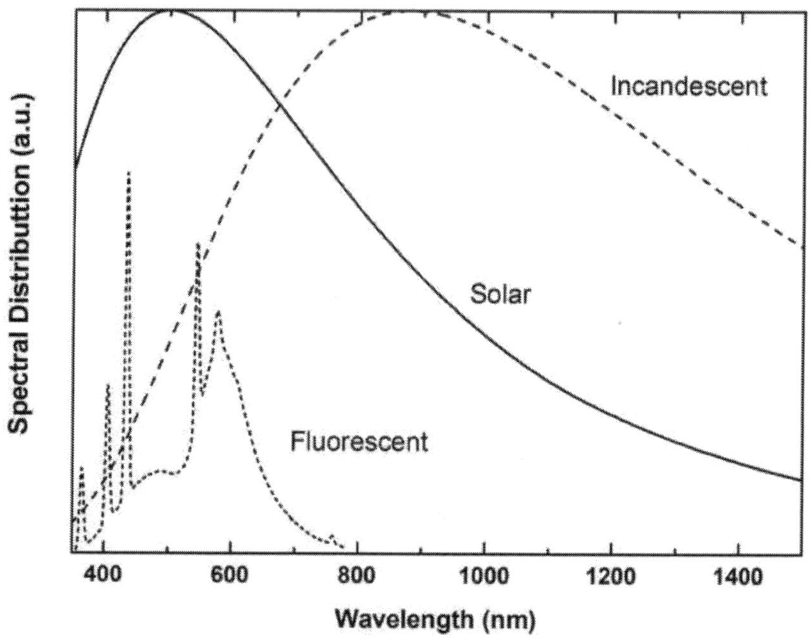

Figure 6 Spectral distribution of solar radiation and emissions from incandescent/fluorescent lamps in the visible and near-infrared wavelength range. Solar and incandescent spectra are calculated from Wien's displacement law at temperatures of 5800 and 3300°K, respectively. Fluorescent spectrum corresponds to a commercially available lamp.

The effects of fluorescent light are particularly deleterious for OOK/ASK modulation formats, in which the electrical power spectrum of OOK/ASK contains a strong DC component. Fluorescent noise cannot be filtered without rejecting a significant portion of the signal in low rate OOK/ASK systems [121], and adaptive filtering [123–125] or spread spectrum [126] techniques are required. Higher rate OOK/ASK systems have a broader spectrum and fluorescent noise may be rejected through high pass filtering, however this leads to a baseline wander effect and additional DC-balanced line coding is required [35]. Low rate PPM systems and their variants (DPIM, DPPM) are less vulnerable to noise from slowly modulated fluorescent lamps than OOK/ASK systems, since the peak spectrums of PPM is located at non-zero frequencies and high-pass filtering rejects fluorescent noise without imposing significant ISI [121]. Rejecting fluorescent light modulated at higher frequencies, however, requires processing techniques similar to those used for OOK/ASK [123]. Finally, SCM and OFDM systems may avoid fluorescent noise altogether by utilizing transmission frequencies that are located outside the fluorescent noise bandwidth.

3.5. VLC systems

With the advent of solid-state lighting based on white LEDs, there is intense interest in the use of visible light for OW communications. Not only do white LEDs for illumination purposes have high energy efficiency and lifetimes, but they also possess significant modulation bandwidth. By modulating the illumination outside of the human perceptual range, significant data rates can be transmitted while satisfying the illumination requirements inside a room [59, 127]. Such VLC channels also possess an SNR that is approximately 30 dB higher than a comparable infrared link since the background interference is largely removed [128]. Moreover, it is not necessary to install additional LEDs for telecommunication purposes. Visible light LEDs are already designed to provide uniform indoor lighting, which means that the effects of shadowing are minimized. In their simplest implementation VLCs only provide broadcast download capability, since the LED lighting will not in general include optical receivers.

VLCs have drawn significant attention from standardization bodies and research institutes in Europe, the USA and Japan. In Japan, VLC standardization efforts have been underway for some time. The VLC consortium [3] was founded in 2003 to investigate and provide a standard platform for VLC systems. In 2007, the VLC standard JEITA CP-1221 was issued by the Japan Electronics and Information Technology Industries Association [129]. European efforts in VLC have also been significant. The OMEGA project [130], funded by the European Commission [131], is tasked to develop the physical channels to enable Gb/s rates for home area networks combining RF, VLC, infrared and power line communications. A total of 20 partners are included in this project which has a budget on the order of €19 million and a European Commission fund of €12.4 million [132]. In the USA, an NSF research fund of $18.5 million has been provided to three universities to establish Smart Lighting Engineering Research Centers, which will develop device and system technologies for VLC communications. In addition, standardization efforts for VLC through the IEEE have been quite active recently [133].

Similar to infrared OW, the transmission rates of VLC systems are limited by the effects of multipath dispersion and noise from the sun and artificial lighting. In principle, the same techniques that have been presented in previous paragraphs to mitigate dispersion and noise may be applied to VLCs. As a result, multiple carrier modulation formats (SCM [134] and OFDM [135, 136]), MIMO systems [137], coding schemes [138, 139], and adaptive equalization methods [140] have been proposed in the literature within the context of VLCs. The transmission rates of VLC systems are also limited by the modulation bandwidth of phosphor-based visible LEDs; the slow time constant of the phosphor they are manufactured with in order to emit yellow light leads to a typical modulation bandwidth of a few MHz. Filtering and equalization at either the transmitter [141] or the receiver [142] alleviates the slow phosphor response, and increases the visible LED bandwidth by an order of magnitude. This bandwidth increase, along with the use of discrete multi-tone OFDM, has enabled the demonstration of VLCs operating at transmission rates that exceed 100 Mb/s [143–145].

An additional aspect of VLCs is that visible LEDS are also required to provide lighting at a constant power level and brightness adjustment options. The human eye is not capable of distinguishing between power variations at the bit level for typical transmission speeds under consideration in VLC systems (tens to hundreds of Mb/s). However, long series of '1's and '0's lead to long-term power variations that are perceived as flickering. Flickering is harmful for the human eye and several approaches, which aim at maintaining a constant DC-component in the emitted VLC signals, have been taken to mitigate its effect. The most common mitigation technique is to use PPM, in which transmitted symbols maintain a pre-determined percentage of '1's and '0's. A performance analysis of PPM variants may be found in [134, 146]. Another approach involves using DC-balanced line coding such as modified 4b5b, 5b6b or 8b10b over OOK [147]. As far as brightness adjustment is concerned, it is accomplished by combining line coding with pulse-width-modulation (PWM) variants [147, 148]. Under this scheme, line coding provides a DC-component whose amplitude can be adjusted by varying the duty cycle of the PWM.

4. OUTDOOR OW SYSTEMS

Another major application area of OW is in outdoor systems. From a commercial perspective, outdoor OW complements optical fibers in LAN interconnections and broadband access networks, providing a high-bandwidth and low-cost alternative [149]. The installation of an outdoor OW system alleviates the most significant expenditure of fiber deployment: the digging up and repairing of roads. In addition, current commercially available systems operate at 10 Gb/s Ethernet speeds, which are comparable to the speeds attained per channel in commercial optical fiber networks.

Directional OW links also have a low probability of intercept and detection compared to broadcast or less-directional RF wireless links. These are desirable attributes for certain applications, especially military communications, where the communicating parties need to remain undetectable by an adversary. In contrast to RF transmissions, where the adversary needs only a low complexity receiver to detect broadcast transmissions or the side-lobes of directional RF antennas, implementation of an OW link in a completely covert fashion requires fully optical pointing, acquisition and tracking without any adjunct RF technology from the third party. Thus, it is essentially impossible for the third party to detect the existence of the link, unless they insert their own receiver into the beam path. The security provided by the optical link is enhanced by means of encryption, which ultimately prevents interception of the actual data being transmitted.

Two key requirements for outdoor OW systems are ever-increasing line speeds and link distances. Multi-Gb/s OW line speeds are feasible at modest distances at the time of writing, owing to technological advances in optical fiber technology and adaptive optics. Fiber networks are expected to advance to 100 Gb/s in the near future, providing the necessary end-point components for OW, as well. As a result, OW line speeds are expected to keep up with their fiber optic counterparts, attaining ultra-high capacities per channel.

In ultra-long range outdoor OW systems, however, the achievable data rates can be severely limited by the earth's atmosphere. The atmosphere has a detrimental impact on optical beams, an impact that worsens as the link length increases [150]. The effects that come into play in making the outdoor OW propagation channel non-ideal are shown schematically in Figure 7. Obscuration (haze, fog, snow, rain) along the path reduces the received power. Even in the absence of significant obscuration, atmospheric turbulence presents significant challenges in the operation of OW links for distances as short as 500 m. Turbulence causes fluctuating intensity at the received power and beam wandering, spreading or breakup. The net result of these effects is manifest as scintillation in the received optical power. Scintillation presents a challenge for the receiver in an OW link, since the receiver must have significant dynamic range, which may need to be more than 40 dB, to handle large variations in the received intensity. Reductions in the received power that take the receiver below the specified threshold correspond to fades. During a deep fade, receiver and background noise dominate over the received signal, rendering the link useless.

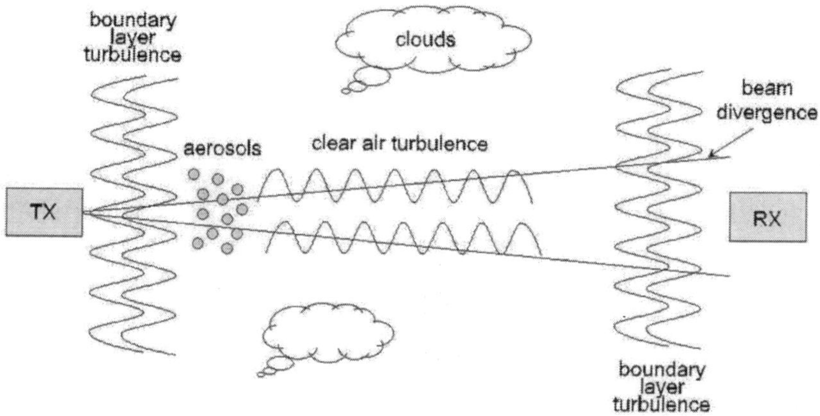

Figure 7 Factors affecting the communication channel for outdoor FSO links.

The stochastic nature of the fades and surges, both in terms of their intensity and their duration, leads to a stochastic description of the channel response. This is in stark contrast with personal communication and indoor systems, where the channel impulse response is modeled as time invariant. Stochastic channel models suitable for outdoor OW are detailed in Section 4.1. Moreover, due to the stochastic nature of the outdoor OW channel, mitigation of channel-imposed degradations is challenging. Applicable mitigation techniques involve some form of diversity, either spatial or temporal, to increase the probability of signal reception. This is accomplished by sending multiple replicas of the original signal over different propagation paths (spatial diversity) or by temporally displacing replicas by an offset that exceeds the fade duration (temporal diversity). Error-correction-coding is also applicable, but in high capacity systems the number of bits affected by a fade increases with the line rate. Still, these techniques can easily add several seconds of latency to a link. In the Gb/s regime, the code length that is required to correct the erroneous bits might be impractically long. Mitigation techniques are discussed in Section 4.2.

The section concludes with the most challenging area of application of outdoor OW systems, satellite communications. These systems employ laser beams that communicate data between terrestrial feeders, satellites and airborne relays and involve link distances that reach up to thousands of kilometers. The impact of atmospheric effects is magnified for such distances, while highly accurate beam acquisition, pointing and tracking systems are required to maintain a working link. OW for satellite communications is presented in Section 4.3.

4.1. Channel modeling

A major effect of the atmosphere on optical beams is the rapid attenuation of their power when they propagate through obscurations (haze, fog, snow and rain). For a single mode Gaussian beam of small angle beam divergence θ and power P_{TX} at the transmitter, the received power at an aperture of area A at range L equals

$$P_{RX} = \frac{2P_{TX}Ae^{-\alpha L}}{\pi\theta^2 L^2}, \tag{4.1}$$

where α is the intensity attenuation coefficient along the obscured path. Typical obscuration losses that are encountered near the ground range from 0.5 dB/km in clear air, 3 dB/km in haze, 50 dB/km in thick fog, and 350 dB/km in impenetrable fog (OW links are more affected by snow and fog than they are by rain). At high altitude in clear air the attenuation may be as low as 0.1 dB/km, although aerosols can still affect the channel [151]. In practice, the exponential decay of the optical power means that OW links do not work over significant distances through obscuration and it is not practical to attempt to operate OW links through significant obscuration by increasing transmitter power. Some additional range can be accomplished by decreasing beam divergence angleθ, although this imposes more rigorous pointing and tracking requirements on the link.

A second performance degradation in outdoor OW systems results from the rapid scintillation of the received power caused by atmospheric turbulence. The local density of the atmosphere is constantly fluctuating because of temperature and pressure fluctuations. When a laser beam propagates through the atmosphere, the randomly varying spatial distribution of refractive index that it encounters causes the following effects [152]:

a) Fluctuating intensity as observed with an optical detector at the end of the path (scintillation).

b) A varying degree of intensity fluctuation that depends on the receiver aperture (the size of the detector, or the size of the receiving optics which direct the collected light to the detector).

c) Progressive deterioration of circularly symmetric Gaussian beams with increasing distance and turbulence strength. The progressive changes that are observed are time-dependent deviations of the beam shape from circular, wander of the centroid of the beam, increase in the width of the beam over and above that expected from diffraction, and breakup of the beam into distinct patches of illumination whose shapes and locations fluctuate with time.

d) A fall in the coherence length of the laser beam.

e) Fluctuations in the angle of arrival of the phase fronts at the receiver.

Theoretical analysis of the effects of the turbulent atmosphere on the propagation of electromagnetic waves has been extensive, and to some extent repetitive, over at least the last 40 years. A substantial amount of this study was carried out in the former Soviet Union, and the books by Tatarski [153] and Chernov [154] are classic accounts of the state of the field at the beginning of the 1960s. Comprehensive reviews of the specific situation regarding laser beam propagation have been given by Ishimaru [155], Strohbehn [156], Zuev [157], Andrews and Phillips [152], and Andrews et al. [158]. There are also a number of other valuable review articles, for example [159, 160]. Many hundreds of additional articles covering the effect of turbulence on propagating laser beams can be found in [152].

The turbulence induced scintillation at the receiver is characterized by the probability density function (pdf) of the received power. The pdf allows the calculation of the probability of fades or surges of a given level below or above the average received power. The pdf can also be used to estimate the fraction of time that a given link will be in a fade or surge state. It is frequently found experimentally that over a wide range of scintillation indices the distribution of log field amplitude is a normal distribution, as are the log intensity variations measured at a point [156]. For log-normal statistics the probability of a fade (or surge) of a given depth can be calculated using the methods described by Yura and McKinley [161], who also provide a means to calculate the mean duration of a fade. An alternative pdf that better describes the distribution of received power in strong turbulence is the γ-γdistribution [157, 159, 162, 163]. If scintillation is reduced by aperture averaging, which is likely for all but very long links, then log-normal statistics are usually adequate to describe the behavior.

The level of scintillation is taken into account in individual pdfs by the scintillation index

$$\sigma_I^2 = \frac{\langle I^2 \rangle - \langle I \rangle^2}{\langle I \rangle^2}. \tag{4.2}$$

I is the intensity measured by a point receiver and angle brackets indicate time averages. In practice, the receiver of an OW link will have a finite collection area and will only act as a point receiver if its diameter D is small enough to satisfy $D << \sqrt{\lambda L}$, where λ is the wavelength used and L is the length of the link. If the receiver diameter is larger than that of a point receiver then the level of scintillation is reduced by aperture averaging and the scintillation index in this case is given by Equation (4.2) with Ireplaced by the received power P.

The strength of turbulence is frequently characterized by the Rytov variance. The Rytov variance is obtained by approximating the solution to Maxwell's equations for electromagnetic wave propagation through a random medium under weak turbulence conditions. For a plane wave, the Rytov variance is calculated as

$$\sigma_{\ln I_R}^2 = \left\langle (\ln I - \langle \ln I \rangle)^2 \right\rangle = 1.23 \cdot C_n^2 k^{7/6} L^{11/6}, \tag{4.3}$$

where C_n^2 is the refractive index structure constant and k is the wavevector magnitude of the light. In weak turbulence, when $\sigma_I^2 < 0.3$, the Rytov approximation is valid and the scintillation index practically coincides with the Rytov variance. As the turbulence strength increases, however, the Rytov approximation no longer provides a precise solution to Maxwell's equations and the Rytov variance increases without limit.

Strong turbulence theory is then required to predict the scintillation index behavior at increasing values of the Rytov variance and it has been shown that the scintillation index saturates at values of the Rytov variance that can range up to 5-6 [158]. Figure 8 show the behavior of the scintillation index, calculated from weak and strong turbulence theory [160], as a function of the Rytov variance.

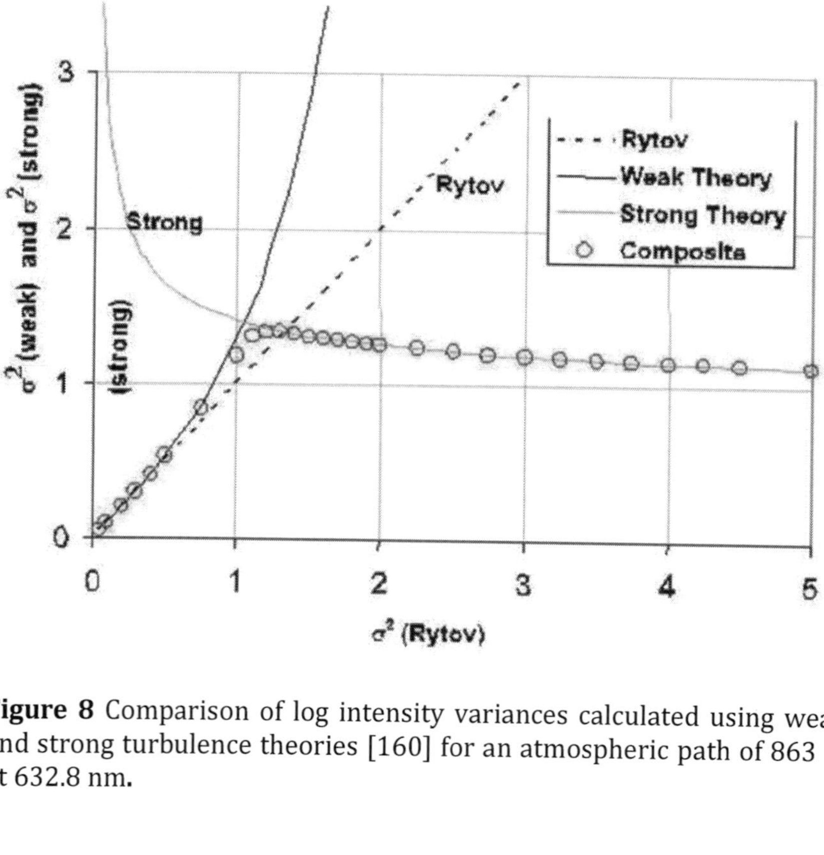

Figure 8 Comparison of log intensity variances calculated using weak and strong turbulence theories [160] for an atmospheric path of 863 m at 632.8 nm.

The aforementioned atmospheric impairments are combined with factors that include (a) the optical power losses in the transmitter and receiver, (b) beam spreading, (c) the Strehl (optical imaging quality) ratio, and (d) pointing errors to calculate link parameters such as the link margin and link availability [164–166]. The link margin is defined as the ratio of the transmitted power to the minimum power that is required to achieve a desired BER at the receiver. Signal fades that exceed the link margin lead to an increased BER for the duration of the fade and extremely deep fades raise the BER to 50%, causing a total loss of data and rendering the link useless. In contrast to other wireless channels, signal fading on the OW channel is relatively slow compared to the data-rate. The duration of a fade might be up to 100 ms or longer when the clock period, which depends on the data rate, is generally less than 20 ns (50 Mb/s). Hence, during a single fade several thousand transmitted bits can be corrupted. The link availability, on the other hand, is measured over a larger timescale and is defined as the percentage of time over a year that the link is available. The link availability is primarily dependent upon the existence of obscurations in the atmosphere. Under heavy obscuration, the atmospheric losses exceed the link margin and, similar to deep fades, the link operates at an increased BER or is unusable for the duration of the obscuration. When link outage resulting from obscuration cannot be tolerated the best solution is to operate a hybrid link where the OW link is operated in parallel with a directional RF link in unlicensed regions of C-band (5.8 GHz) or Ku-band. Hybrid OW/RF systems are detailed in the next section.

4.2. Fade mitigation techniques

(1) There are four principal methods for mitigating the effects of fading at the physical layer: aperture averaging, diversity, error correction, and adaptive optics.

Aperture averaging: Aperture averaging reduces the scintillation index by using a receiver whose aperture diameter *D* is larger than the lateral correlation distance on the linkλL−−−$\sqrt{\sqrt{\lambda L}}$. Measurements of the effect of the aperture size on scintillation index have been given in [167]. These measurements, made in intermediate strength turbulence, are in reasonable agreement with the weak turbulence formula given in [152]

$$F = \frac{\sigma_I^2(D)}{\sigma_I^2(0)} = \left[1 + 1.07 \cdot \left(\frac{kD^2}{4L}\right)\right]^{-1}, \tag{4.4}$$

which describes the reduction *F* in the scintillation index for a receiver of diameter *D* compared to a point receiver and holds for plane waves with a small inner scale l_0 satisfying $l_0 << \sqrt{L/k}$. The aperture averaging factor does not depend on the exact shape of the aperture and depends primarily on receiver aperture area [168].

(2) Diversity techniques: Diversity techniques use multiple identical transmissions from sources to receivers [169, 170]. The multiple

transmissions take uncorrelated paths and as a result the effective level of scintillation and the probability of a fade can be reduced. The most common diversity approach uses multiple transmitter and receiver apertures (MIMO), which are spaced further apart than the lateral correlation distance on the link [171–173]. If n independent apertures are used and if the received signals are logically combined then the probability of a fade of a given depth is reduced to prob $(P \leq P_{\min})^n$, with $P_{\min}$ being the minimum power at each receiver aperture. For a fair comparison of the diversity approach the probability should be evaluated for the same total power transmitted, which means that the fade reduction is really $\mathrm{frac}\left(\frac{P_{Total}}{n} \leq P_{\min}\right)^n$, where frac($x$) denotes the fractional part of x.

An alternative approach to spatial diversity is to use time-delayed diversity (TDD), described in [174, 175]. In this approach, n total transmissions of the same data take place between a single transmitter and a single receiver. The replicated transmissions are delayed by an offset τ which is longer than expected fade duration. The replicas are re-synchronized, buffered and logically combined at the receiver. If the separate transmissions of data are truly independent then the BER is reduced from its single channel value of BER to BER^n. In practice, the exact reduction in BER must be calculated from the joint pdf for the signal received at different delay times, which requires knowledge of the temporal covariance matrix. Experimental measurements [175] have shown that in weak turbulence a delay of about 20 ms provides decorrelation between successive transmissions. The required decorrelation time becomes shorter as turbulence gets stronger, which is precisely when diversity approaches of all types become most valuable.

(3) Error correction: In some scenarios and especially for airborne terminals, size, power and weight constraints may not allow for multiple transmissions and therefore diversity cannot be implemented. Additionally, diversity systems can get very complex because of issues such as optics alignment, signal synchronization, interference effects, and others. Error correction is a simpler fade mitigation technique that has been shown to give satisfactory results [176] on SISO (single input single output) channels. Forward-error-correction (FEC), in particular, adds additional bit patterns to each coded packet being transmitted, which allows all transmitted bits to be recovered if some are lost due to fading. For example, in a (255,223) Reed-Solomon code that uses 8-bit symbols, there are 255 codeword bytes. Each codeword contains 223 bytes of data and 32 bytes for parity checking. The decoder can correct

any 16 symbol errors in the code word: errors up to 16 bytes anywhere in the codeword can be automatically corrected. Still, this method only works well if fades are of short duration. In this example, the fade duration should be limited to 128-bit periods. Unfortunately, the transmission time of a code word in a high data rate OW system is much shorter than even the shortest fades. At 1.25 Gb/s, which is a commercially available OW data rate, 255 bytes takes only 1.6 μs for transmission. As a result, improved coding and interleaving techniques are being investigated for application in higher rate OW systems.

Recent study has suggested that BER improvement can be obtained by the use of low density parity check (LDPC) codes [163]. It is not clear, however, that in this work the temporal characteristics of the strongly fading channel were properly accounted for. The performance of a link-layer LDPC staircase FEC system (these codes are called IRA - Irregular Repeat-Accumulate codes [177]) has also been tested over a Fast-Ethernet OW link [176]. In the presented test scenario the mentioned code was able to overcome all packet losses (11% of lost data). Another approach is to employ interleaving in parallel with coding. Interleaving provides a temporal diversity by spacing coded symbols by time intervals that are longer than expected fade durations. As a result, a fade only affects a single symbol per word in an interleaved in system and the length of the required code is drastically reduced. Further details about coding and interleaving for OW channels can be found in the specialized literature [177–180].

The drawback of FEC, and especially interleaving, is increased latency. The data have to be buffered at the transmitter and the receiver for at least the duration of the used coding constraint length. A rule of thumb suggests using block lengths that have a constraint length of several times the maximum expected fade duration. The processing time needed for encoding and decoding must also be added to the total processing delay. In general, the overall transmission delay can be expected to exceed 100 ms and therefore it is not practical to implement FEC-interleaving schemes for delay-critical data. Latency can be alleviated by introducing rate-compatible (or rateless) coding into the error correction scheme [181]. A key aspect of rateless coding is that the receiver is capable of decoding the original information, with high probability, even if only part of the original information is available. To this end, the transmitter creates a sequence of frames based on the original data, and the receiver combines the collected frames and decodes the original data once enough frames have been received.

(4) Adaptive optics: In adaptive optics (AO) a deformable mirror is used to pre-distort a transmitted laser beam so that on its passage through the turbulent atmosphere it arrives at the receiver as an undistorted wave. In most long range OW demonstrations that have been carried out to-date AO has been used to attempt to mitigate the effects of the turbulent channel. Unfortunately, as pointed out by Stotts et al. [164] "the ability of an adaptive optics (AO) system to compensate turbulence along a path is limited by the transmitter and receiver Rayleigh ranges, proportional to the diameter of the optics squared and inverse of the wavelength of light utilized." AO is ineffective when the total range is greater than the sum of the Rayleigh ranges. The Rayleigh range is $R_R = 0.7D^2/\lambda$, where D is the diameter of the transmitter or receiver aperture, and λ is the wavelength. For example with a 1.55-μm transmitter, and a 100-mm diameter receiver R_R = 4.5 km. An AO receiver can correct the wavefronts that actually arrive at a receiver aperture, for example to couple the light into single mode fiber. But, full reciprocal AO that couples transmitter and receiver is not possible unless the range is sufficiently short, or when the transmitter and receiver are not moving fast laterally with respect to each other. Although AO has had considerable success in optical astronomy, its complexity, cost, and inability to correct for long range effects makes its widespread adoption for OW problematic. A full discussion of AO is beyond the scope of this article, but the study of Fried and co-workers [182–187] is particularly important in this context.

4.3. Satellite links and high-altitude platforms

Outdoor OW links are a high-bandwidth and low-power-consumption alternative in satellite communications [4] and OW links have been successfully demonstrated in satellite-to-satellite, earth-to-satellite and earth-to-moving-vehicle links [188–199]. Satellite systems are the most challenging category of outdoor OW systems, because of the very long distances that the optical beams propagate through vacuum or the atmosphere. Optical links that operate in vacuum (inter-satellite and space communications) are at least tens of thousands of kms long. Even if the vacuum is considered to be ideal (lossless), the associated path losses that increase with the square of the link distance are enormous, and vary from approximately 150 dB in satellite-to-satellite communications to approximately 230 dB in Mars-to-Earth communications. Earth-to-satellite optical links involves shorter distances, up to approximately 36,000 km for a geostationary orbit, but the total system losses are increased due to scattering and absorption in the atmosphere [150, 152]. Hence, a key consideration in satellite communications is the received optical power and several techniques have been proposed to mitigate the transmission associated losses. Optical amplification is an appealing candidate [200–203], especially in earth-to-satellite uplinks where low power consumption limitations do not apply. Another possible candidate in earth-to-satellite communications is the use of flying vehicles and high-altitude-platforms as beam relays [204–206]. When no amplification or intermediate relays are available, as in deep space communications [207], the losses may not be mitigated and more sophisticated receivers that operate at very low powers (photon-counting) are required [208].

A second important aspect of satellite systems is the selection of suitable modulation formats and coding schemes, depending on the application scope. Power efficiency is critical in satellite communications, since satellites recharge only by means of solar energy. Recently, a 5.6-Gb/s inter-satellite link demonstrated the suitability of binary phase shift keying (BPSK) for satellite OW links [192]. Deep space communications, on the other hand, are plagued by extreme path losses and favor the deployment of modulation formats with high peak-to-average power ratios, which ensure that energy peaks are detected. Multilevel PPM systems have been shown to be suitable for deployment in deep space communications [150, 209]. Efficient error-correcting coding is also required in deep space communications due to the extremely low powers received, and concatenated convolutional codes have being shown to perform close to the theoretical capacity limit [209]. In earth-to-satellite systems, coherent phase modulation schemes provide high sensitivity and the use of BPSK has been previously reported on a similar 5.6 Gb/s 142 km OW link [210]. PPM has also been studied under the prism of providing sufficient peak power to penetrate through clouds [211, 212]. Error-correcting codes help to mitigate deep atmospheric fades, but there is a high possibility that the OW link will be lost due to adverse weather conditions altogether [213]. In such cases, temporal or spatial diversity is required to maintain a working earth-to-satellite optical link [214, 215].

A final challenge in satellite communications is beam acquisition, pointing, and tracking. Longer links typically require more accurate and sophisticated acquisition, pointing and tracking systems to maintain an operational OW link, since the slightest error is magnified by the link distance. From a communication perspective, pointing and tracking are not only important for maintaining the optical link, but also contribute to the link power penalty since misalignment caused by pointing and tracking errors result in power fluctuations at the receiver [216–218]. Pointing and tracking errors are caused mainly by (a) satellite vibrations [218], (b) errors in the calculation of the point-ahead angle [4], and (c) beam wander effects [219]. Pointing and tracking errors may be reduced by using advanced automated control methods in the pointing and tracking element [4], while their impact may be dampened by allowing adaptive variations in the transmitted power [218] or a broader optical beam [220]. In earth-to-satellite communications pointing and tracking also suffer from fades in the atmosphere, and it is possible that re-acquisition of the beam is required after a fade, leading to extended capacity loss. Reference [221] discusses how inertial sensors can be utilized to counter-effect beam loss due to atmospheric fading.

5. HYBRID OW/RF SYSTEMS

OW links can support high data rate communications in large unlicensed bandwidths, but their data rates and reliability are severely affected by atmospheric phenomena that include absorption, scattering and atmospheric turbulence [158]. While in clear conditions atmospheric attenuation for an OW link can be as low as a fraction of a few dB/km, heavy snow or fog can increase this attenuation to hundreds of dBs/km. Atmospheric turbulence, particularly in hot dry climates, causes signal fading that can degrade the link performance significantly. Besides, when the links involve mobility, maintaining accurate pointing and tracking over long distances becomes highly challenging, and link deterioration or complete link failure occurs. A key requirement is therefore to increase the reliability of OW links.

Improving the reliability of OW links, while still exploiting their high data carrying capabilities, is achieved by operating the OW link in conjunction with an RF link. Under this scheme of operation, the RF part may be used for link control, maintenance and backup functions that require relatively low line rates, while the broadband OW part transfers the bulk of data. This hybrid operation scheme improves reliability since RF links can be much less sensitive to atmospheric conditions and pointing errors, depending on their frequency of operation. In addition, when the OW links fail due to obscurations or fading, the RF links can act as a backup still providing a significant percentage of the OW system bandwidth. The RF data rate can be increased over a given bandwidth using higher order modulations, for instance quadrature amplitude modulation at the expense of transmitted power and system complexity. Even though RF frequencies have regulation restrictions and large bandwidths for high data rates at desired frequencies are difficult to obtain, unlicensed bands at 2.4, 5, and 60 GHz are available for the implementation of hybrid OW/RF systems.

The increased reliability of hybrid OW/RF over OW-only links enables them as a low-cost rapidly deployable last-mile communication technology. In last-mile applications, OW/RF can provide connectivity over longer distances within the eye safety limitations of laser radiation. In addition, OW/RF links can be used for both fixed as well as mobile multigigabit backbone networks [222–224]. Such networks may include ship-to-shore, ship-to-ship, ground-to-aircraft, and aircraft-to-aircraft links. The use of hybrid OW/RF systems for mobile robot team applications has been discussed in [225]. Taking advantage of the high bandwidths of OW links and reliability of RF links, such robot teams can transmit real-time video data in applications such as search and rescue operations. The use of hybrid OW/RF systems can also facilitate RF spectrum planning and utilization. If OW/RF hybrid systems are designed to meet a required quality of service then some of the existing RF-only links can be replaced with hybrid links. Hybrid links typically utilize only a fraction of the RF bandwidth in comparison with RF-only links [226]. The released RF spectrum can subsequently be made available for more critical applications.

5.1. Hybrid OW/RF channel model

In a hybrid OW/RF system, it is necessary to include both the OW and RF channel impairments in the overall system performance. In general, OW and RF links, depending on the frequency of operation, can be modeled independently to include the weather effects. As a result, separate channel models for the OW and RF links are mostly adequate to predict the hybrid system performance. Most practical OW systems employ intensity modulation with direct detection (IM/DD). The OW channel propagation results in a multiplicative effect on the transmitted optical intensity of the laser source with a channel gain factor. The channel gain includes atmospheric attenuation, pointing error effects and atmospheric turbulence. A detailed discussion of the channel impact on OW only systems has been presented in Section 4. Since the OW link operates with direct line-of-sight (LOS), the RF link too has LOS, and therefore the RF link can be modeled as a Ricean channel and often as an additive white Gaussian noise (AWGN) channel for good channel conditions. However, for longer and high altitude links severe fading in the RF link can occur. One example is [222], where the link operates over a distance of 150 km at a height of 10,000 ft. Both frequency-flat and frequency-selective fading were observed for a system operating in the X- and Ku-band at a data rate of 274 Mb/s.

When higher RF frequencies are considered (such as at millimeter wavelengths) the OW and RF channels may not be truly independent. Higher RF frequency bands support higher data rates, but they are also more affected by atmospheric effects and pointing errors. Consequently, higher RF frequencies are affected by scintillations, and moderate to heavy rain events were reported to affect both OW laser link at 850 nm and millimeter wave RF [227]. Pointing errors can also affect scintillation statistics for both channels [228]. In such a case, pointing error correlations between the two links cannot be ignored.

5.2. Hybrid OW/RF link architecture and networking

The RF link in hybrid OW/RF channels can be used for link acquisition, data transmission control and/or data transfer. Accordingly, hybrid systems are divided into three categories, depending on the usage of the RF link, as follows:

(1) Hybrid OW/RF using RF as the control link: The RF link is generally more reliable than the OW link and less sensitive to pointing. Thus, it can be used to assist in the pointing and acquisition of the OW link. It can also be used to provide packet retransmission requests. Successful RF retransmission requests can significantly improve system performance, as observed in [222] for a multi-Gb/s data rate system.

(2) Hybrid OW/RF with switching and data splitting: Since the OW link is the primary channel, the RF link can be used either as a backup or as a parallel link carrying a fraction of the transmitted data. An experimental 2.9 km hybrid OW/RF link is described in [229], where the RF link works in a standby mode. The laser link provides connection at 155 Mb/s and the received signal level is checked every 5 s. When the signal falls below a certain threshold, data transmission switches to the RF link, which operates at 2.4 GHz with a data rate of 11 Mb/s. Data transmission continues on the RF link until an improved signal level is detected and then the transmission switches back to the laser link.

Although this hybrid OW/RF experiment is important, this approach under-utilizes resources, as only one link is active at any given time.

When using both OW and RF links concurrently, data is partitioned between the links. In [224], a dynamic load switching algorithm is employed and incremental load shifting from one link to the other is carried out, based on the status of the two links. For this purpose, BER is recorded on both links every minute and the averaging effect over a window is used for switching. A practical coded OW/RF switching scheme using an automatic load distribution between the links is described in [230]. The data packets are encoded using Raptor codes and sent over the two links at their respective rates of 1 Gb/s and 96 Mb/s. Both the links employ cyclic redundancy check (CRC) codes and discard corrupted packets. The receiver decodes the entire message when there are a sufficient number of correctly received packets. The transmitter continues sending encoded packets for a given message until it receives a 1-bit feedback message. Although for a given reliability level the data rate of this approach is less due to Raptor code overhead, the average data rate is improved by nearly six times in some cases. The use of Raptor codes for hybrid OW/RF systems is also considered in [231].

(3)Hybrid OW/RF with codeword partitioning: If the transmitted data are encoded for error correction, separate encoding and decoding over OW and RF links do not fully exploit the available diversity. For example, when the OW link deteriorates, a data packet encoded and decoded only over the OW link may still be received in error. If, however, some of the encoded bits are transmitted over a more reliable RF link, there is a higher probability of correctly receiving the packet. This idea has been recently explored by many researchers. In general, as more and more bits are transmitted over a more reliable RF channel, the hybrid link remains operational even with less available OW SNR. This gives rise to trade-offs between OW and RF link parameters.

A non-uniform rate-compatible low density parity check (LDPC) code is proposed in [232] for partitioning data into OW and RF links. The code rate is changed through puncturing of the coded bits. To get insights into the behavior of LDPC codes over hybrid channels, density evolution for hybrid channels is described in [226]. For a given code rate and OW/RF signal-to-noise ratios, the minimum RF data rate needed for the convergence of the decoder algorithm is obtained. This helps in the code requirements and design of hybrid systems. For example, the receiver can ask for an increase or decrease in RF rate in steps. Hybrid channel density evolution analysis for irregular LDPC codes shows that some further improvement in performance can be obtained. The performance improvement is higher for stronger turbulence channels. An experimental testbed verification demonstrating the joint decoding of LDPC codes over the two channels is carried out in [233]. Convolutional and turbo codes for hybrid channels are also investigated in the literature [234, 235]. Since these codes require simpler encoders than decoders, they can be useful at transmitters with limited resources. The convolutionally coded bits are partitioned into OW and RF channels using an optimized pattern and interleaved before modulation [234]. This bit-interleaved coded modulation is able to exploit the diversity of the two links. In [235], error analysis for turbo codes in hybrid channels showing trade-offs between RF and OW SNRs is described. Optimal turbo-coded bit allocations for puncturing and RF transmission are analyzed and related algorithms are also presented.

From a networking perspective, hybrid OW/RF networks are required to provide high-rate data connectivity with low end-to-end delay. A multihop network with short length links will provide high availability since the shorter OW links are less vulnerable to outages [224]. However, adding more hybrid nodes is costly and it also increases end-to-end delay. A hybrid wireless network of backbone nodes is considered in [223], where the nodes can be pointed to new nodes if necessary. Topology discovery, dissemination, link restoration, optimal OW/RF configurations and routing are very critical for network operation. These functions are accomplished through cooperation of the RF and OW subsystems. For link restoration, an increase in OW transmit power, bit rate reduction and hybrid link switching were explored. Alternatively, OW beam search along with RF connectivity can be used. If the OW link fails in a switched hybrid system then the connection switches to the RF link and optical beam probing can be continued until the OW link is restored [223]. A detailed OW/RF network analysis, with links using codeword partitioning, is an open research problem.

6. CONCLUSIONS AND FUTURE DIRECTIONS

This article has presented an overview of OW technologies, emphasizing their deployment in communication systems. Four distinct OW system categories (personal communication, indoor, outdoor, and hybrid OW/RF) have been discussed and key requirements and major applications areas have been identified for each category. Personal communication OW systems are required to provide high rate connectivity and very short connection establishment times. Both have been implemented in the IrDA-standardized physical layer and protocol stack. The current state of the art has enabled data transmission at Gb/s speeds and this technology is currently under standardization by the IrDA Giga-IR group. Future standardization efforts include a technology merge of infrared and USB technologies (IrUSB), as well as a comprehensive analysis and optimization of available protocols with the new Gb/s speeds in mind.

Indoor OW systems are required to allow for user mobility, which can be quite challenging due to the dispersive nature of the indoor channel. Dispersion mitigation is of critical importance in indoor OW and suitable mitigation techniques, including multi-spot diffusing links and coding, have been presented in this article. The deployment of OW has been limited in the past years, but recently VLCs have emerged as a future high-potential indoor OW technology. VLCs are being actively investigated by the VLCs Consortium and the IEEE 802.15.7 Task Group. The VLCs Consortium has already produced standards for VLCs and is currently cooperating with IrDA to promote OW systems for personal and indoor communications.

Outdoor OW systems are required to provide high data rates and remain functional at link distances that range from hundreds of meters to thousands of kms. Outdoor systems are seriously affected by atmospheric impairments, mainly obscuration and fading. These effects are increasingly deleterious as the link length increases and mitigation techniques like temporal/spatial diversity, coding/interleaving and aperture averaging are required to maintain link operation. At present, 10 Gb/s Ethernet compliant outdoor OW systems are commercially available for metro applications from several vendors. In the near future, link speeds are expected to rise, following the bandwidth increase in high-speed optical components that currently operate at 40-100 Gb/s.

Finally, hybrid OW/RF systems have been introduced to augment outdoor OW systems with increased reliability. The inclusion of the RF subsystem facilitates link establishment and operation, enhances error correction and data transfer control, and provides a backup link in the event of a failure. Hybrid systems are currently under research and researchers aim to identify which cooperation schemes between the OW and RF links provide the most reliable overall system operation.

COMPETING INTERESTS

The authors declare that they have no competing interests.

REFERENCES

1. Gfeller F, Bapst U: Wireless in-house data communication via diffuse infrared radiation. Proc IEEE 1979, 67(11):1474-1486.

2. Infrared Data Association (IrDA)[http://www.irda.org]

3. Visible Light Communications Consortium[http://www.vlcc.net]

4. Chan VWS: Optical satellite networks. IEEE/OSA J Lightw Technol 2003, 21(11):2811-2827. 10.1109/JLT.2003.819534

5. IEEE Standard for Wireless LAN Medium Access Control (MAC) and Physical Layer (PHY) Specifications. 1997.

6. IEEE 802.15 Working Group for WPAN[http://www.ieee802.org/15/]

7. IrDA: Advanced Infrared (AIr) MAC Draft Protocol Specification--Version 1.0, Infrared Data Association. 1999.

8. IrDA: Advanced Infrared Logical Link Control (AIrLC) Specification--Version 0.1, Infrared Data Association. 1999.

9. IrDA: Advanced Infrared Physical Layer Specification (AIr-PHY) --Version 1.0, Infrared Data Association. 1998.

10. IrDA: Point and Shoot Profile--Version 1.1, Infrared Data Association. 2000.

11. IrDA: Physical Layer Specification--Version 1.4, Infrared Data Association. 2001.

12. IrDA: Serial Infrared Physical Layer Specification--Giga-IR Addition Version 1.0, Infrared Data Association. 2009.

13. IEC 60825-1, Safety of Laser Products--Part 1: equipment classification, requirements, and user's guide, edition 1.2 International Electrotechnical Commission 2001.

14. IrDA: Serial Infrared Link Access Protocol (IrLAP) --Version 1.1, Infrared Data Association. 1996.

15. IrDA: Link Management Protocol (IrLMP)--Version 1.1, Infrared Data Association. 1996.

16. IrDA: Tiny TP: a flow-control mechanism for use with IrLMP--Version 1.1, Infrared Data Association. 1996.

17. IrDA: Object Exchange Protocol (OBEX™)--Version 1.5, Infrared Data Association. 2009.

18. IrDA: Link Access Protocol Specification for Giga-IR Addition--Version 1.0, Infrared Data Association. 2009.

19. IrDA: IrLAP Fast Connect (Application Note)--Version 1.0, Infrared Data Association. 2002.

20. IrDA: Minimal IrDA Protocol Implementation (IrDA Lite)--Version 1.0, Infrared Data Association. 1996.

21. IrDA: IrSimple (Infrared Simple) Profile--Version 1.01, Infrared Data Association. 2007.

22. IrDA: Serial Infrared Sequence Management Protocol for IrSimple--Version 1.0, Infrared Data Association. 2005.

23. IrDA: Infrared Financial Messaging (IrFM™) Point and Pay Profile--Version 1.0, Infrared Data Association. 2002.

24. IrDA: 'IrCOMM': Serial and Parallel Port Emulation over IR (Wire Replacement)--Version 1.0, Infrared Data Association. 1995.

25. IrDA: Infrared Universal Bus Specification--Version 0.11, Infrared Data Association. 2010.

26. Boucouvalas AC: Indoor ambient light noise and its effect on wireless optical links. IEE Proc Optoelectron 1996, 143(6):334-338. 10.1049/ip-opt:19960881

27. Kahn J, Barry J: Wireless infrared communications. Proc IEEE 1997, 85(2):265-298. 10.1109/5.554222

28. Kahn JM, Krause WJ, Carruthers JB: Experimental characterization of non-directed indoor infrared channels. IEEE Trans Commun 1995, 43(234):1613-1623.

29. Pérez-Jiménez R, Rabadán JA, López-Hernández FJ: Filtered modulation schemes for short distance infrared wireless communications.IEEE Trans Consumer Electron 2000, 46(2):275-282. 10.1109/30.846658

30. Valadas RT, Tavares AR, de Oliveira Duarte AM, Moreira AC, Lomba CT: The infrared physical layer of the IEEE 802.11 Standard for wireless local area networks. IEEE Commun Mag 1998, 36(12):107-112. 10.1109/35.735887

31. Shiu D, Kahn JM: Differential pulse position modulation for power-efficient optical communication. IEEE Trans Commun 1999, 47(8):1201-1210. 10.1109/26.780456

32. Ghassemlooy Z, Hayes AR: Digital pulse interval modulation for IR communication systems--a review. Int J Commun Sys 2000, 13(7-8):519-536. 10.1002/1099-1131(200011/12)13:7/8<519::AID-DAC454>3.0.CO;2-5MATH

33. Hirt W, Hassner M, Heise N: IrDA-VFIr (16 Mb/s): modulation code and system design. IEEE Personal Commun 2001, 8(1):58-71. 10.1109/98.904900

34. Widmer AX, Franaszek PA: A DC-balanced, partitioned-block, 8B/10B transmission code. IBM J Res Dev 1983, 27(5):440-451.

35. Street AM, Samaras K, O'Brien DC, Edwards DJ: Closed form expressions for baseline wander effects in wireless IR applications. IEE Electron Lett 1997, 33(12):1060-1062. 10.1049/el:19970739

36. Ghassemlooy Z: Investigation of the baseline wander effect on indoor optical wireless system employing digital pulse interval modulation.IET Commun 2008, 2(1):53-60. 10.1049/iet-com:20050552

37. Carruthers JB: JM Kahn, Multiple-subcarrier modulation for non-directed wireless infrared communication. IEEE J Sel Areas Commun1996, 14(3):538-546. 10.1109/49.490239

38. You R, Kahn JM: Average power reduction techniques for multiple-subcarrier intensity-modulated optical signals. IEEE Trans Commun2001, 49(12):2164-2171. 10.1109/26.974263MATH

39. Gonzalez O, Perez-Jimenez R, Rodriguez S, Rabadan J, Ayala A: OFDM over indoor wireless optical channel. IEE Proc Optoelectron2005, 152(4):199-204. 10.1049/ip-opt:20045065

40. Mesleh R, Elgala H, Haas H: On the performance of different OFDM based optical wireless communication systems. OSA J Opt Commun Netw 2011, 3: 620-628.

41. Armstrong J: OFDM for optical communications. IEEE/OSA J Lightw Technol 2009, 27(3):189-2041.

42. Chatzimisios P, Boucouvalas AC: Packet delay analysis of the advanced infrared (AIr) CSMA/CA MAC protocol in optical wireless LANs. Int J Commun Syst 2005, 18(3):307-331. 10.1002/dac.705

43. Vitsas V, Boucouvalas AC: Performance analysis of the advanced infrared (AIr) CSMA/CA MAC protocol for wireless LANs. Wirel Netw2003, 9(5):495-507. 10.1023/A:1024692217327

44. Barker P, Vitsas V, Boucouvalas AC: Simulation analysis of advanced infrared (Alr) MAC wireless communications protocol. IEE Proc Circ Dev Syst 2002, 149(3):193-197. 10.1049/ip-cds:20020444

45. Bianchi G: Performance analysis of the IEEE 802.11 distributed coordination function. IEEE J Sel Areas Commun 2000, 18(3):535-547. 10.1109/49.840210

46. Crow BP, Widjaja I, Kim LG, Sakai PT: IEEE 802.11 wireless local area networks. IEEE Commun Mag 1997, 35(9):116-126. 10.1109/35.620533

47. Huang P, Chatzimisios P, Boucouvalas A: Optimizing IrDA throughput by including processing time with physical layer consideration. OSA J Opt Netw 2005, 4: 323-331. 10. 1364/JON.4.000323

48. Vitsas V, Boucouvalas AC: Optimization of IrDA IrLAP link access protocol. IEEE Trans Wirel Commun 2003, 2(5):926-938. 10.1109/TWC.2003.816776

49. Boucouvalas AC, Huang Pi: Modelling and optimising TinyTP over IrDA stacks. EURASIP J Wirel Commun Netw 2005, 2005(1):45-56.

50. Boucouvalas AC, Huang Pi: OBEX over IrDA: performance analysis and optimization by considering multiple applications. IEEE/ACM Trans Netw 2006, 14(6):1292-1301.

51. Deccio CT, Ekstrom J, Partridge DR, Tew KB, Knutson CD: A study of the suitability of IrOBEX for high-speed exchange of large data objects. In Proc IEEE GLOBECOM. Volume 5. San Francisco; 2003:2664-2668.

52. Shah AM, Ara SS, Kitazumi G, Matsumoto M: IrSimple Modeling and Performance Evaluation for High-Speed Infrared Communications. InProc IEEE GLOBECOM. San Francisco; 2006:1-6.

53. IrDA: Burst Data Protocol (IrBurst)--Version 1.0, Infrared Data Association. 2004.

54. Alam MS, Shawkat SA, Gontaro K, Mitsuji M: IrBurst modeling and performance evaluation for large data block exchange over high-speed IrDA links. IEICE Trans Commun 2008, 91(1):274-285.

55. Huang P, Boucouvalas A: IrBurst modelling and performance analysis in the presence of transmission errors. Wirel Personal Commun2007, 41(1):111-125. 10.1007/s11277-006-9131-9

56. Huang P, Boucouvalas A: Future personal "e-payment": IRFM. IEEE Wirel Commun 2006, 13(1):60-66. 10.1109/MWC.2006.1593526

57. Chu T-S, Gans M: High speed infrared local wireless communication. IEEE Commun Mag 1987, 25(8):4-10.

58. Heatley DJT, Wisely DR, Neild I, Cochrane P: Optical wireless: the story so far. IEEE Commun Mag 1998, 36(12):72-74. 79-82 10.1109/35.735881

59. Pang G, Kwan T, Liu H, Chan C-H: LED wireless. IEEE Ind Appl Mag 2002, 8(1):21-28. 10.1109/2943.974354

60. Hranilovic S: Wireless Optical Communication Systems. Springer, New York, NY; 2004.

61. Barry JR, Kahn JM: Link design for nondirected wireless infrared communications. OSA Appl Opt 1995, 34: 3764-3776.

62. Wang K, Nirmalathas A, Lim C, Skafidas E: 4 × 12.5 Gb/s WDM optical wireless communication system for indoor applications. IEEE/OSA J Lightw Technol 2011, 29(13):1988-1996.

63. Pakravan MR, Simova E, Kavehrad M: Holographic diffusers for indoor infrared communication systems. Int J Wirel Inf Netw 1997, 4(4):259-274. 10.1023/A:1018876326494

64. Elgala H, Mesleh R, Haas H: Indoor broadcasting via white LEDs and OFDM. IEEE Trans Consumer Electron 2009, 55(3):1127-1134.

65. Jungnickel V, Forck A, Haustein T, Kruger U, Pohl V, von Helmolt C: Electronic tracking for wireless infrared communications. IEEE Trans Wirel Commun 2003, 2(5):989-999. 10.1109/TWC. 2003.817419

66. Plant DV, Kirk AG: Optical interconnects at the chip and board level: challenges and solutions. Proc IEEE 2000, 88(6):806-818. 10.1109/5.867694

67. Kim G, Han X, Chen R: An 8-Gb/s optical backplane bus based on microchannel interconnects: design, fabrication, and performance measurements. IEEE/OSA J Lightw Technol 2000, 18(11):1477-1486. 10.1109/50.896207

68. Bevan MG, Darrin MAG, Walts SC, Schneider W, Mills CS, Conde RF: Free-space optical data bus for spacecraft. In Proceedings of the 3rd annual Earth Science Technology Conference. Maryland, USA; 2003.

69. O'Brien DC, Faulkner GE, Zyambo EB, Jim K, Edwards DJ, Stavrinou P, Parry G, Bellon J, Sibley MJ, Lalithambika VA, Joyner VM, Samsudin RJ, Holburn DM, Mears RJ: Integrated transceivers for optical wireless communications. IEEE J Sel Topics Quantum Electron2005, 11(1):173-183.

70. Miller D: Device requirements for optical interconnects to silicon chips. Proc IEEE 2009, 97(7):1166-1185.

71. Shelby RM, Hoffnagle JA, Burr GW, Jefferson CM, Bernal M-P, Coufal H, Grygier RK, Gunther H, Macfarlane RM, Sincerbox GT: Pixel-matched holographic data storage with megabit pages. OSA Opt Lett 1997, 22(19):1509-1511.

72. Burr GW, Ashley J, Coufal H, Grygier RK, Hoffnagle JA, Jefferson CM, Marcus B: Modulation coding for pixel-matched holographic data storage. OSA Opt Lett 1997, 22(9):639-641.

73. Bisaillon E, Brosseau DF, Yamamoto T, Mony M, Bernier E, Goodwill D, Plant DV, Kirk AG: Free-space optical link with spatial redundancy for misalignment tolerance. IEEE Photon Technol Lett 2002, 14(2):242-244.

74. Hranilovic S, Kschischang FR: Short-range wireless optical communication using pixelated transmitters and imaging receivers. InProceedings of the IEEE International Conference on Communications. Volume 2. Paris, France; 2004:891-895.

75. Hranilovic S, Kschischang FR: A pixelated MIMO wireless optical communication system. IEEE J Sel Topics Quantum Electron 2006, 12(4):859-874.

76. Mesleh R, Elgala H, Haas H: Optical spatial modulation. OSA J Opt Commun Netw 2011, 3(3):234-244.

77. Takase D, Ohtsuki T: Spatial multiplexing in optical wireless MIMO communications over indoor environment. IEICE Trans Commun 2006, E89-B(4):1364-1370. 10.1093/ietcom/e89-b.4.1364

78. Villan R, Voloshynovskiy S, Koval O, Pun T: Multilevel 2-D bar codes: toward high-capacity storage modules for multimedia security and management. IEEE Trans Inf Foren Sec 2006, 1(4):405-420.

79. Mohamed MDA, Hranilovic S: Two-dimensional binary halftoned optical intensity channels. IET Commun 2008, 2(1):11-17. 10.1049/iet-com:20060385

80. Siegel PH: Information-theoretic limits of two-dimensional optical recording channels. In Proceedings of the IEEE Optical Data Storage Topical Meeting. Montreal, QC, Canada; 2006:165-167.

81. Huang L, Mathew G, Chong TC: Channel modeling and target design for two-dimensional optical storage systems. IEEE Trans Mag 2005, 41(8):2414-2424.

82. Joseph J, Waldman DA: Homogenized Fourier transform holographic data storage using phase spatial light modulators and methods for recovery of data from the phase image. Appl Opt 2006, 45: 6374-6380. 10.1364/AO.45.006374

83. Pu A, Psaltis D: High-density recording in photopolymer-based holographic three-dimensional disks. OSA Appl Opt 1996, 35(14):2389-2398.

84. Takase D, Ohtsuki T: Optical wireless MIMO communications (OMIMO). In IEEE Global Telecom Conf. Volume 2. Dallas; 2004:928-932.

85. Faucher J, Venditti MB, Plant DV: Application of parallel forward-error correction in two-dimensional optical-data links. IEEE/OSA J Lightw Technol 2003, 21(2):466-475. 10.1109/JLT.2003.808762

86. Ohbuchi E, Hanaizumi H, Hock LA: Barcode readers using the camera device in mobile phones. In Int Conference on Cyberworlds. Tokyo, Japan; 2004:260-265.

87. Kahn JM, Barry JR, Audeh MD, Carruthers JB, Krause WJ, Marsh GW: Non-directed infrared links for high-capacity wireless LANs. IEEE Personal Commun 1994, 1(2):12-25.

88. Kotzin M, van den Heuvel A: A duplex infra-red system for inbuilding communications. In Proceedings of the 36th IEEE Vehicular Technology Conference. Volume 36. Dallas, Texas; 1986:179-185.

89. Barry JR, Kahn JM, Krause WJ, Lee EA, Messerschmitt DG: Simulation of multipath impulse response for indoor wireless optical channels.IEEE J Sel Areas Commun 1993, 11(3):367-379. 10.1109/49.219552

90. Alqudah YA, Kavehrad M: MIMO characterization of indoor wireless optical link using a diffuse-transmission configuration. IEEE Trans Commun 2003, 51(9):1554-1560. 10.1109/TCOMM.2003.816945

91. Hashemi H, Yun G, Kavehrad M, Behbahani F, Galko PA: Indoor propagation measurements at infrared frequencies for wireless local area networks applications. IEEE Trans Veh Technol 1994, 43(3):562-576. 10.1109/25.312790

92. Pakravan MR, Kavehrad M: Indoor wireless infrared channel characterization by measurements. IEEE Trans Veh Technol 2001, 50(4):1053-1073. 10.1109/25.938580

93. Carruthers JB, Kahn JM: Modeling of nondirected wireless infrared channels. IEEE Trans Commun 1997, 45(10):1260-1268. 10.1109/26.634690

94. Jungnickel V, Pohl V, Nonnig S, von Helmolt C: A physical model of the wireless infrared communication channel. IEEE J Sel Areas Commun 2002, 20(3):631-640. 10.1109/49.995522

95. Lee H-S: A photon modeling method for the characterization of indoor optical wireless communication. Prog Electromagn Res PIER 2009, 92: 121-136.

96. Lopez-Hernandez FJ, Perez-Jimenez R, Santamarıa A: Ray-tracing algorithms for fast calculation of the channel impulse response on diffuse IR wireless indoor channels. Opt Eng 2000, 39(10):2775-2780. 10.1117/1.1287397

97. Gonzalez O, Rodriguez S, Perez-Jimenez R, Mendoza BR, Ayala A: Error analysis of the simulated impulse response on indoor wireless optical channels using a Monte Carlo-based ray-tracing algorithm. IEEE Trans Commun 2005, 53(1):124-130. 10.1109/ TCOMM. 2004.840625

98. González O, Rodríguez S, Pérez-Jiménez R, Mendoza BR, Ayala A: Comparison of Monte Carlo ray-tracing and photon-tracing methods for calculation of the impulse response on indoor wireless optical channels. OSA Opt Express 2011, 19: 1997-2005.

99. Cocheril Y, Vauzelle R: A new ray-tracing based wave propagation model including rough surfaces scattering. Prog Electromagn Res PIER2007, 75: 357-381.

100. Audeh MD, Kahn JM, Barry JR: Performance of pulse position modulation on measured non-directed indoor infrared channels. IEEE Trans Commun 1996, 44(6):654-659. 10.1109/26.506380

101. Audeh MD, Kahn JM, Barry JR: Decision-feedback equalization of pulse-position modulation on measured nondirected indoor infrared channels. IEEE Trans Commun 1999, 47(4):500-503. 10.1109/26.764921

102. Lee DC, Kahn JM, Audeh MD: Trellis-coded pulse position modulation for indoor wireless infrared communications. IEEE Trans Commun1997, 45(9):1080-1087. 10.1109/26.623072

103. Lee DC, Kahn JM: Coding and equalization for PPM on wireless infrared channels. IEEE Trans Commun 1999, 47(2):255-260. 10.1109/26.752131

104. Gfeller F, Hirt W: Advanced infrared (AIr): physical layer for reliable transmission and medium access. In Proceedings of the International Zurich Seminar on Broadband Communications. Zurich, Switzerland; 2000:77-84.

105. Lee D, Kahn J: Experimental 25-Mb/s wireless infrared link using 4-PPM with scalar decision-feedback equalization. In Proceedings of the IEEE International Conference on Communications. Volume 1. Atlanta, GA, USA; 1998:26-30.

106. Marsh GW, Kahn JM: 50-Mb/s diffuse infrared free-space link using on-off keying with decision-feedback equalization. IEEE Photon Technol Lett 1994, 6(10):1268-1270.

107. Marsh GW, Kahn JM: Performance evaluation of experimental 50-Mb/s diffuse infrared wireless link using on-off keying with decision-feedback equalization. IEEE Trans Commun 1996, 44(11):1496-1504. 10.1109/26.544466

108. Yun G, Kavehrad M: Indoor infrared wireless communications using spot diffusing and fly-eye receivers. Can J Electron Comput E 1993, 18(4):151-157.

109. Alsaadi FE, Elmirghani JMH: Mobile multigigabit indoor optical wireless systems employing multibeam power adaptation and imaging diversity receivers. OSA J Opt Commun Netw 2011, 3: 27-39.

110. Alsaadi FE, Elmirghani JMH: High-speed spot diffusing mobile optical wireless system employing beam angle and power

adaptation and imaging receivers. IEEE/OSA J Lightw Technol 2010, 28(16):2191-2206.

111. Jivkova ST, Hristov BA, Kavehrad M: Power-efficient multispot-diffuse multiple-input-multiple-output approach to broad-band optical wireless communications. IEEE Trans Veh Technol 2004, 53(3):882-889. 10.1109/TVT.2004.825779

112. Jivkova ST, Kavehrad M: Multispot diffusing configuration for wireless infrared access. IEEE Trans Commun 2000, 48(6):970-978. 10.1109/26.848558

113. Jivkova ST, Kavehrad M: Holographic optical receiver front end for wireless infrared indoor communications. OSA Appl Opt 2001, 40: 2828-2835.

114. Kahn JM, You R, Djahani P, Weisbin AG, Teik BK, Tang A: Imaging diversity receivers for high-speed infrared wireless communication.IEEE Commun Mag Vol 1998, 36: 88-94.

115. Carruthers J, Kahn JM: Angle diversity for nondirected wireless infrared communication. IEEE Trans Commun 2000, 48(6):960-969. 10.1109/26.848557

116. Jivkova S, Kavehrad M: Receiver designs and channel characterization for multi-spot high-bit-rate wireless infrared communications. IEEE Trans Commun 2001, 49(12):2145-2153. 10.1109/26.974261

117. Kavehrad M, Jivkova S: Indoor broadband optical wireless communications: optical subsystems design and their impact on channel characteristics. IEEE Wirel Commun Mag Vol 2003, 10: 30-35.

118. Al-Ghamdi AG, Elmirghani JMH: Line strip spot-diffusing transmitter configuration for optical wireless systems influenced by background noise and multipath dispersion. IEEE Trans Commun 2004, 52(1):37-45. 10.1109/TCOMM.2003.822160

119. Al-Ghamdi AG, Elmirghani JMH: Multiple spot diffusing geometries for indoor optical wireless communication systems. Int J Commun Syst2003, 16: 909-922. 10.1002/dac.627

120. Jungnickel V, von Helmolt C, Kruger U: Broadband wireless infrared LAN architecture compatible with ethernet protocol. Electron Lett1998, 34(25):2371-2372. 10.1049/el:19981676

121. Narasimhan R, Audeh MD, Kahn JM: Effect of electronic-ballast fluorescent lighting on wireless infrared links. IEE Proc Optoelectron 1996, 143(6):347-354. 10.1049/ip-opt:19960877

122. Moreira AJC, Valadas RT, de Olveira Duarte AM: Optical interference produced by artificial light. Wirel Netw 1997, 3(2):131-140. 10.1023/A:1019140814049

123. Fernando XN, Krishnan S, Sun H, Kamyar KM: Adaptive denoising at infrared wireless receivers. Proc SPIE 2003, 5074: 199-207.

124. Rajbhandari S, Ghassemlooy Z, Angelova M: Effective denoising and adaptive equalization of indoor optical wireless channel with artificial light using the discrete wavelet transform and artificial neural network. IEEE/OSA J Lightw Technol 2009, 27(20):4493-4500.

125. Rajbhandari S, Ghassemlooy Z, Angelova M: Wavelet--artificial neural network receiver for indoor optical wireless communications.IEEE/OSA J Lightw Technol 2011, 29(17):2651-2659.

126. Wong KK, O'Farrell T: Spread spectrum techniques for indoor wireless IR communications. IEEE Wirel Commun 2003, 10(2):54-63. 10.1109/MWC.2003.1196403

127. Komine T, Nakagawa M: Fundamental analysis for visible-light communication system using LED lights. IEEE Trans Consumer Electron2004, 50(1):100-107. 10.1109/TCE.2004.1277847

128. O'Brien D, Katz M: Optical wireless communications within fourth-generation wireless systems [invited]. OSA J Opt Netw 2005, 4(6):312-322. 10.1364/JON.4.000312

129. Japan Electronics and Information Technology Industries Association[http://www.jeita.or.jp]

130. OMEGA, the Home Gigabit Access Project[http://www.ict-omega.eu/]

131. The OMEGA Project Flyer [Online]. Available: http://www.ict-omega.eu/fileadmin/documents/flyers/OMEGA_Project_Flyer.pdf

132. Future networks, Projects Portfolio, Seventh EU Framework Programme for Research and Technological Development, Information Society and Media, European Commission 2008.

133. IEEE 802.15 WPAN-Visible Light Communication Interest Group (IGvlc)[http://ieee802.org/15/pub/IGvlc.html]

134. Sugiyama H, Haruyama S, Nakagawa M: Experimental investigation of modulation method for visible-light communications. IEICE Trans Commun 2006, E89-B(12):3393-3400. 10.1093/ietcom/e89-b.12.3393

135. Elgala H, Mesleh R, Haas H, Pricope B: OFDM visible light wireless communication based on white LEDs. In Proc IEEE Vehicular Technol Conf. Volume 4212879. Dublin, Ireland; 2007:2185-2189.

136. Grubor J, Randel S, Langer K-D, Walewski JW: Broadband information broadcasting using LED-based interior lighting. IEEE/OSA J Lightw Technol 2008, 26(24):3883-3892.

137. Zeng L, O'Brien DC, Minh HL, Faulkner GE, Lee K, Jung D, Oh Y, Won T: High data rate multiple input multiple output (MIMO) optical wireless communications using white led lighting. IEEE J Sel Areas Commun 2009, 27(9):1654-1662.

138. Kwon JK: Inverse source coding for dimming in visible light communications using NRZ-OOK on reliable links. IEEE Photon Technol Lett2010, 22(19):1455-1457.

139. Sunghwan K, Sung-Yoon J: Novel FEC coding scheme for dimmable visible light communication based on the modified Reed-Muller codes.IEEE Photon Technol Lett 2011, 23(20):1514-1516.

140. Komine T, Lee JH, Haruyama S, Nakagawa M: Adaptive equalization system for visible light wireless communication utilizing multiple white led lighting equipment. IEEE Trans Wirel Commun 2009, 8(6):2892-2900.

141. Minh HL, O'Brien D, Faulkner G, Zeng L, Lee K, Jung D, Oh Y: 80 Mbit/s visible light communications using pre-equalized white LED.

InProc Eur Conf Opt Commun no. P.6.06. Brussels, Belgium; 2008:1-2.

142. Minh HL, O'Brien D, Faulkner G, Zeng L, Lee K, Jung D, Oh Y, Won ET: 100-Mb/s NRZ visible light communications using a postequalized white LED. IEEE Photon Technol Lett 2009, 21(15):1063-1065.

143. Grubor J, Lee SCJ, Langer K-D, Koonen T, Walewski JW: Wireless high-speed data transmission with phosphorescent white-light LEDs. InProc Eur Conf Opt Commun, PD3.6. Berlin, Germany; 2007:1-2.

144. Vučić J, Kottke C, Nerreter S, Langer K-D, Walewski JW: 513 Mbit/s visible light communications link based on DMT-modulation of a white LED. IEEE/OSA J Lightw Technol 2010, 28(24):3512-3518.

145. Vučić J, Kottke C, Nerreter S, Büttner A, Langer K-D, Walewski JW: White light wireless transmission at 200+ Mb/s net data rate by use of discrete-multitone modulation. IEEE Photon Technol Lett 2009, 21(20):1511-1513.

146. Lee K, Park H: Modulations for visible light communications with dimming control. IEEE Photon Technol Lett 2011, 23(16):1136-1138.

147. IEEE 802.15.7 VLC PHY/MAC Proposal [https://mentor.ieee.org/802.15/dcn/09/15-09-0733-00-0007-vlc-phy-mac-proposal-samsung-etri.pdf]

148. Ntogari G, Kamalakis T, Walewski JW, Sphicopoulos T: Combining illumination dimming based on pulse-width modulation with visible-light communications based on discrete multitone. OSA J Opt Commun Netw 2011, 3: 56-65.

149. Liu Q, Qiao C, Mitchell G, Stanton S: Optical wireless communication networks for first- and last-mile broadband access [Invited]. OSA J Opt Netw 2005, 4: 807-828. 10.1364/JON.4.000807

150. Chan VWS: Free-space optical communications. IEEE/OSA J Lightw Technol 2006, 24(12):4750-4762.

151. Zilberman A, Kopeika NS: Laser beam wander in the atmosphere: implications for optical turbulence vertical profile sensing with imaging LIDAR. J Appl Remote Sens 2008., 2(023540):

152. Andrews LC, Phillips RL: Laser Beam Propagation Through Random Media. 2nd edition. SPIE Press, Bellingham, WA; 2005.

153. Tatarski VI: Wave Propagation in a Turbulent Medium. McGraw-Hill, New York; 1961. (Reprinted by Dover, New York, 1967)MATH

154. Chernov LA: Wave Propagation in a Random Medium. McGraw-Hill, New York; 1960. (Reprinted by Dover, New York, 1967)

155. Ishimaru AK: Wave Propagation and Scattering in Random Media. Volume I and II. Academic Press, New York; 1978. Vol. II Chapters 16-20 are especially relevant

156. Strohbehn JW: Laser Beam Propagation Through the Atmosphere. Springer-Verlag, New York; 1978.

157. Zuev VE: Laser Beams in the Atmosphere. Consultants Bureau, New York; 1982.

158. Andrews LC, Phillips RL, Hopen CY: Laser Beam Scintillation with Applications. SPIE Press, Bellingham, WA; 2001.

159. Lawrence RS, Strohbehn JW: A survey of clear-air propagation effects relevant to optical communications. Proc IEEE 1970, 58: 1523-1545.

160. Fante RW: Electromagnetic beam propagation in turbulent media. Proc IEEE 1975, 63: 1669-1692.

161. Yura HT, McKinley WG: Optical scintillation statistics for IR ground-to-space laser communication systems. OSA Appl Opt 1983, 22: 3353-3358.

162. Al-Habash MA, Andrews LC, Phillips RL: Mathematical model for the irradiance probability density function of a laser beam propagating through turbulent media. Opt Eng 2001, 40(8):1554-1562. 10.1117/1.1386641

163. Anguita JA, Djordjevic IB, Neifeld MA, Vasic BA: Shannon capacities and error-correction codes for optical atmospheric

turbulence channels. OSA J Opt Netw 2005, 4(9):586-601. 10.1364/JON.4.000586

164. Stotts LB, Kolodzy P, Pike A, Graves B, Daugherty D, Douglas J: Free space optical communication link budget estimation. OSA Appl Opt2010, 49(28):5333-5343.

165. Arnon S: Effects of atmospheric turbulence and building sway on optical wireless-communication systems. OSA Opt Lett 2003, 28: 129-131.

166. Kedar D, Arnon S: Optical wireless communication through fog in the presence of pointing errors. OSA Appl Opt 2003, 42: 4946-4954.

167. Yuksel H, Milner S, Davis CC: Aperture averaging for optimizing receiver design and system performance on free-space optical communication links. OSA J Opt Netw 2005, 4(8):462-475. 10.1364/JON.4.000462

168. Yuksel H, Davis CC: Aperture averaging analysis and aperture shape invariance of received scintillation in free space optical communication links. In Proc SPIE Free-Space Laser Commun VI. Volume 6304. San Diego; 2006.

169. Wilson SG, Brandt-Pearce M, Cao Q, Baedke M: Optical repetition MIMO transmission with multi-pulse PPM. IEEE J Sel Areas Commun2005, 23(9):1901-1910.

170. Navidpour SM, Uysal M, Kavehrad M: BER performance of free-space optical transmission with spatial diversity. IEEE Trans Wirel Commun2007, 6(8):2813-2819.

171. Goldsmith A, Jafar SA, Jindal N, Vishwanath S: Capacity limits of MIMO channels. IEEE J Sel Areas Commun 2003, 21(5):684-702. 10.1109/JSAC.2003.810294

172. Khalighi M-A, Schwartz N, Aitamer N, Bourennane S: Fading reduction by aperture averaging and spatial diversity in optical wireless systems. OSA J Opt Commun Netw 2009, 1: 580-593.

173. García-Zambrana A, Castillo-Vázquez C, Castillo-Vázquez B: Outage performance of MIMO FSO links over strong turbulence and

misalignment fading channels. OSA Opt Express 2011, 19: 13480-13496.

174. Davis CC, Smolyaninov II: The effect of atmospheric turbulence on bit-error-rate in an on-off-keyed optical wireless system. Proc SPIE Free-Space Laser Commun Laser Imag 2002, 4489: 126-137.

175. Trisno S, Smolyaninov II, Milner SD, Davis CC: Delayed diversity for fade resistance in optical wireless communication system through simulated turbulence. Proc SPIE Opt Transmission Syst Equip WDM Netw III 2005, 5596: 385-393. Philadelphia, Pennsylvania

176. Henniger H, Gonzalez A: Transmission scheme and error protection for simplex long-distance atmospheric FSO systems. Special Issue Mediterranean J Electron Commun Hybrid RF Opt Wirel Commun 2006, 2(3):118-126.

177. Zhu X, Kahn JM: Performance bounds for coded free-space optical communications through atmospheric turbulence channels. IEEE Trans Commun 2003, 51: 1233-1239. 10.1109/TCOMM.2003.815052

178. Djordjevic IB: LDPC-coded MIMO optical communication over the atmospheric turbulence channel using Q-ary pulse-position modulation.OSA Opt Express 2007, 15: 10026-10032.

179. Greco JA: Design of the high-speed framing, FEC, and interleaving hardware used in a 5.4 km free-space optical communication experiment. SPIE Proc Free-space Laser Communications IX 2009., 7464:

180. Henniger H, David F, Giggenbach D, Rapp C: Evaluation of FEC for the atmospheric optical IM/DD channel. SPIE Proc Free-Space Laser Communication Technologies XV 2003., 4975:

181. Anguita JA, Neifeld MA, Hildner B, Vasic B: Rateless coding on experimental temporally correlated FSO channels. IEEE/OSA J Lightw Technol 2010, 28(7):990-1002.

182. Fried DL: Limiting resolution looking down through atmosphere. JOSA 1966, 56: 1380-1384. 10.1364/JOSA.56.001380

183. Fried DL: Aperture averaging of scintillation. JOSA 1967, 57: 169-172. 10.1364/JOSA.57.000169

184. Fried DL, Seidman JB: Laser-beam scintillation in the atmosphere. JOSA 1967, 57: 181-185. 10.1364/JOSA.57.000181

185. Greenwood DP, Fried DL: Power spectra requirements for wave-front-compensative systems. JOSA 1976, 66: 193-206. 10.1364/JOSA.66.000193

186. Fried DL: Anisoplanatism in adaptive optics. JOSA 1982, 72: 52-61. 10.1364/JOSA.72.000052

187. Barchers JD, Fried DL: Optimal control of laser beams for propagation through a turbulent medium. JOSA 2002, A: 1779-1793.

188. Sodnik Z, Furch B, Lutz H: Optical intersatellite communication. IEEE J Sel Topics Quantum Electron 2010, 16(5):1051-1057.

189. Tolker-Nielsen T, Oppenhaeuser GT: In orbit test result of an operational optical intersatellite link between ARTEMIS and SPOT4, SILEX.Proc SPIE 2002., 6635(59430):

190. Jono T, Takayama Y, Kura N, Ohinata K, Koyama Y, Shiratama K, Sodnik Z, Demelenne B, Bird A, Arai K: OICETS on-orbit laser communication experiments. Proc SPIE 2006, 6105(610503):1-11.

191. Fujiwara Y, Jono T, Yamawaki T, Arai K, Toyoshima M, Kunimori H, Sodnik Z, Bird A, Demelenne B: Optical inter-orbit communications engineering test satellite (OICETS). Acta Astronautica 2007, 61(1-6):163-175. 10.1016/j.actaastro.2007.01.021

192. Smutny B, Kampfner H, Muhlnikel G, Sterr U, Wandernoth B, Heine F, Hildebrand U, Dallmann D, Reinhardt M, Freier A, Lange R, Bohmer K, Feldhaus T, Muller J, Weichert A, Seel S, Meyer R, Czichy R: 5.6 Gb/s optical intersatellite communication link. Proc SPIE 2009., 7199(719906):

193. Wilson KE, Lesh JR, Yan TY: GOPEX: a laser uplink to the Galileo spacecraft on its way to Jupiter. Proc SPIE 1993, 1866: 138-146.

194. Wilson KE, Lesh JR, Araki K, Arimoto Y: Overview of the ground-to-orbit lasercom demonstration (GOLD). Proc SPIE 1997, 2990: 23.

195. Toyoshima M, Yamakawa S, Yamawaki T, Arai K, Reyes M, Alonso A, Sodnik Z, Demelenne B: Ground-to-satellite optical link tests between Japanese laser communications terminal and European geostationary satellite ARTEMIS. Proc SPIE 2004, 5338: 1-15.

196. Horwath J, Perlot N, Knapek M, Moll F: Experimental verification of optical backhaul links for high-altitude platform networks: atmospheric turbulence and downlink availability. Int J Satell Commun Netw 2007, 25(5):501-528. 10.1002/sat.888

197. Perlot N, Knapek M, Giggenbach D, Horwath J, Brechtelsbauer M, Takayama Y, Jono T: Results of the optical downlink experiment KIODO from OICETS satellite to optical ground station oberpfaffenhofen (OGS-OP). Proc SPIE 2007., 6457(645704):

198. Horwath J, Fuchs C: Aircraft to ground unidirectional laser-comm terminal for high resolution sensors. Proc SPIE 2009., 7199(719909):

199. Baister G, Kudielka K, Dreischer T, Tüchler M: Results from the DOLCE (deep space optical link communications experiment) Project. Proc SPIE 2009., 7199(71990B):

200. Pe'er IL, Naftali N, Yogev A: High power, solar pumped, Nd:YAG, laser amplifier for free space laser communication. Proc SPIE 1997, 3139: 194-204.

201. Rochat E, Dändliker R, Haroud K, Czichy RH, Roth U, Costantini D, Holzner R: Fiber amplifiers for coherent space communication. IEEE J Sel Topics Quantum Electron 2001, 7(1):64-81. 10.1109/2944.924012

202. Polishuk A, Arnon S: Optimization of a laser satellite communication system with an optical preamplifier. J Opt Soc Am A 2004, 21(7):1307-1315. 10.1364/JOSAA.21.001307

203. Dawson JW, Messerly MJ, Beach RJ, MYSEA Stappaerts, Sridharan AK, Pax PH, Heebner JE, Siders CW, Barty CPJ: Analysis of the scalability of diffraction-limited fiber lasers and amplifiers to high average power. OSA Opt Express 2008, 16(17):13240-13266.

204. Djuknic GM, Freidenfelds J, Okunev Y: Establishing wireless communications services via high-altitude aeronautical platforms: a

concept whose time has come? IEEE Commun. Mag 1997, 35(9):128-135.

205. Knapek M, Horwath J, Moll F, Epple B, Courville N, Bischl H, Giggenbach D: Optical high-capacity satellite downlinks via high-altitude platform relays. Proc SPIE 2007., 6709(67090E):

206. Fidler F, Knapek M, Horwath J, Leeb WR: Optical communications for high-altitude platforms. IEEE J Sel Topics Quantum Electron 2010, 16(5):1058-1070.

207. Boroson DM, Biswas A, Edwards BL: MLCD: Overview of NASA's Mars laser communications demonstration system. Proc SPIE 2004, 5338: 16-28.

208. Mendenhall JA, Candell LM, Hopman PI, Zogbi G, Boroson DM, Caplan DO, Digenis CJ, Hearn DR, Shoup RC: Design of an optical photon counting array receiver system for deep-space communications. Proc IEEE 2007, 95(10):2059-2069.

209. Moision B, Hamkins J: Deep-space optical communications downlink budget: modulation and coding. IPN Prog Rep 2003, 42-154: 1-28.

210. Lange R, Smutny B, Wandernoth B, Czichy R, Giggenbach D: 142 km, 5.625 Gb/s free-space optical link based on homodyne BPSK modulation. Proc SPIE 2006., 6105(61050A):

211. Muhammad SS, Javornik T, Jelovcan I, Ghassemlooy Z, Leitgeb E: Comparison of hard-decision and soft-decision channel coded M-ary PPM performance over free space optical links. Eur Trans Telecommun 2008, 20(8):746-757.

212. Djordjevic IB, Vasic B, Neifeld MA: Multilevel coding in free-space optical MIMO transmission with Q-ary PPM over the atmospheric turbulence channel. IEEE Photon Technol Lett 2006, 18(14):1491-1493.

213. Toyoshima M, Takayama Y, Takahashi T, Suzuki K, Kimura S, Takizawa K, Kuri T, Klaus W, Toyoda M, Kunimori H, Jono T, Arai K: Ground-to-satellite laser communication experiments. IEEE Aerospace Electron Syst Mag 2008, 23(8):10-18.

214. Zhu X, Kahn JM: Free-space optical communication through atmospheric turbulence channels. IEEE Trans Commun 2002, 50(8):1293-1300. 10.1109/TCOMM.2002.800829

215. Lee J, Chan VWS: Diversity coherent and incoherent receivers for free-space optical communication in the presence and absence of interference. OSA J Opt Commun Netw 2009, 1(5):463-483.

216. Chen CC, Gardner CS: Impact of random pointing and tracking errors on the design of coherent and incoherent optical intersatellite communication links. IEEE Trans Commun 1989, 37(3):252-260. 10.1109/26.20099

217. Liu X: Optimal transmitter power of an intersatellite optical communication system with reciprocal Pareto fading. OSA Appl Opt 2010, 49: 915-919.

218. Arnon S, Kopeika NS, Kedar D, Zilberman A, Arbel D, Livne A, Guelman M, Orenstain M, Michalik H, Ginati A: Performance limitation of laser satellite communication due to vibrations and atmospheric turbulence: down-link scenario. Int J Satell Commun Netw 2003, 21(6):561-573. 10.1002/sat.769

219. Perlot N: Atmospheric occultation of optical intersatellite links: coherence loss and related parameters. OSA Appl Opt 2009, 48(12):2290-2302.

220. Arnon S, Rotman S, Kopeika NS: Beam width and transmitter power adaptive to tracking system performance for free-space optical communication. OSA Appl Opt 1997, 36(24):6095-6101.

221. Ortiz GG, Lee S, Monacos S, Wright M, Biswas A: Design and development of a robust ATP subsystem for the altair UAV-to-ground lasercomm 2.5 Gb/s demonstration. Proc SPIE 2003, 4975: 103-114.

222. Stotts LB, Andrews LC, Cherry PC, Foshee JJ, Kolodzy PJ, Mcintire WK, Northcott M, Phillips RL, Pike HA, Stadler B, Young DW: Hybrid optical RF airborne communications. Proc IEEE 2009, 97(6):1109-1127.

223. Milner SD, Davis CC: Hybrid free space optical/RF networks for tactical operations. In Proc IEEE Milcom 1. Monterey; 2004:409-415.

224. Izadpanah H, Elbatt T, Kukshya V, Dolezal F, Ryu BK: High-availability free space optical and RF hybrid wireless networks. IEEE Wirel Commun 2003, 10(2):45-53. 10.1109/MWC.2003.1196402

225. Derenick J, Thorne C, Spletzer S: On the deployment of a hybrid FSO/RF mobile ad-hoc network. In Proc IEEE/RSJ Int Conf Intelligent Robots and Syst. Edmonton; 2005:3990-3996.

226. Tapse H, Borah DK: Hybrid Optical/RF channels: characterization and performance study using low density parity check codes. IEEE Trans Commun 2009, 57: 3288-3297.

227. Kukshya V, Rappaport TS, Izadpanah H, Tangonan G, Guerrero RA, Mendoza JK, Lee B: Free-space optics and high-speed RF for next generation networks--propagation measurements. In Proc IEEE VTC-Fall 1. Vancouver; 2002:616-620.

228. Borah DK, Voelz DG: Pointing error effects on free space optical communication links in the presence of atmospheric turbulence. IEEE/OSA J Lightwave Technol 2009, 27: 3965-3973.

229. Akbulut A, Ilk HG, Ari F: Design, availability and reliability analysis on an experimental outdoor FSO/RF communication system. Proc Transparent Opt Netw 2005, 1: 403-406.

230. Zhang W, Hranilovic S, Shi C: Soft-switching hybrid FSO/RF links using short-length raptor codes: design and implementation. IEEE J Sel Areas Commun 2009, 27: 1698-1708.

231. Abdul Hussein A, Oka A, Nguyen T, Lampe L: Rateless coding for hybrid free-space optical and radio-frequency communication. IEEE Trans Wirel Commun 2010, 9: 907-913.

232. Vangala S, Pishro-Nik H: A highly reliable FSO/RF communication system using efficient codes. In Proc IEEE Globecom. Washington, D.C.; 2007:2232-2236.

233. Luna R, Borah DK, Jonnalagadda R, Voelz D: Experimental demonstration of a hybrid link for mitigating atmospheric turbulence effects in free space optical communication. IEEE Photon Technol Lett 2009, 21: 1196-1198.

234. He B, Schober R: Bit-interleaved coded modulation for hybrid RF/FSO systems. IEEE Trans Commun 2009, 57: 3753-3763.

235. Tapse H, Borah DK: J Perez-Ramirez, Hybrid optical/RF channel performance analysis for turbo codes. IEEE Trans Commun 2011, 59: 1389-1399.

3

BIO-INSPIRED PHOTONIC STRUCTURES: PROTOTYPES, FABRICATIONS AND DEVICES

Feng Liu[1], Biqin Dong[2] and Xiaohan Liu[3]

[1] Laboratory of Opto-electrical Material and Device, Department of Physics, Shanghai Normal University, Shanghai, China

[2] Department of Mechanical Engineering, Northwestern University, Evanston, USA

[3] Department of Physics, Fudan University, Shanghai, China

1. INTRODUCTION

Like the ability of electron regulation of electronic semiconductors, the photonic analogs usually considered as photonic structure materials are regarded as essential for light manipulation [1, 2]. With particular designs of photonic structures, they are expected to achieve different far-field and near-field optical features and thus lead to a perspective in all-optical circuit [3]. Though humankind has entered the nano-scale realm several decades ago, it is still a hard task for engineers to explore novel optical functional devices due to the limited experiences and originalities on artificial photonic structures design and the desired optical features. Additionally, it is also great challenges to fabricate photonic structures, owing to their sub-optical-wavelength to sub-micron featured sizes, especially in a high dimensional way by nano-fabrication technologies today.

By contrast, nature are found to develop photonic structures millions of years before our initial attempts. Diversified photonic structures, most of which are sophisticated and hierarchic, are revealed in beetles, butterflies, sea animals and even plants in recent surveys [4–10]. The exhibited optical features are regarded to have particular biological functions such as signal communications, conspecific recognition, and camouflage, which are optimized under selection pressure. Naturally, the occurring photonic structures provide us ideal 'blueprints' on design and stimulate similar optical functional devices. Various fabrication methods of bio-inspired photonic structures are explored [11–16]. By chemical methods (e.g. Sol-Gel, colloidal crystallization, chemical systhesis), nanoimprint lithography and nanocasting, physical layer deposition (PLD), atomic layer deposition (ALD), and etc., bio-inspired photonic structures, their reverse counterparts, and applications are achieving greater success than ever before.

This Chapter will review the typical bio-inspired photonic structures and focus on the biomimetic fabrications, the corresponding optical functions and the prototypes of optical devices. It is organized as follows: four subsections are introduced in Section 2, in which each describes one catagory of bio-inspired optical functional devices. In every subsection, the nature prototype is introduced first, then followed by biomimetic fabrication methods and optical features of artificial analogs, finally closed on the bio-inspired optical devices. A brief perspective is given in Section 3.

2. BIO-INSPIRED OPTICAL FUNCTIONAL DEVICES

2.1. Anti-Reflection Devices

2.1.1. Prototypes

In arthropodal animals such as butterflies, nipple arrays with typical spacing of optical wavelength or subwavelength are commonly found on surfaces of their compound eyes, which are believed helpful to the light-harvest efficiency of the biological visual system [17]. With optical impedance matching to the ambience, the light transmission are enhanced. Another analogous examples are the transparent wings of some lepidoptera insects like hawkmoths [18] and cicada [19] (Fig.1 (a)). The tapered pillars lead to gradual changes of refraction index in view of effective medium theory (Fig.1 (b)) andtherefore play a key role in minimizing the reflection over broadband and large view angles. In order to physically explain the anti-reflection origin, the Fresnel equations are given as follows.

$$r_s = \left|\frac{n_1 cosi_1 - n_2 cosi_2}{n_1 cosi_1 + n_2 cosi_2}\right|, \quad (1)$$

$$r_p = \left|\frac{n_1 cosi_2 - n_2 cosi_1)}{(n_1 cosi_2 + n_2 cosi_1}\right|, \quad (2)$$

where *rs* and *rp* refer to reflection coefficients of *s* polarised light (the electric field of the light perpendicular to the incident plane) and *p* polarised light (the electric field in the incident light plane),n_1 and n_2 are refraction indices of neighboring mediums, respectively. The relationship between incident angle i_1 and refraction angle i_2 is given by Snell's Law $n_1 \sin i_1 = n_2 \sin i_2$. From the equations, e.g., for normal incidence, *rs* and *rp* are suppressed to near zero if n_1 and n_2 have very close values [20, 21].

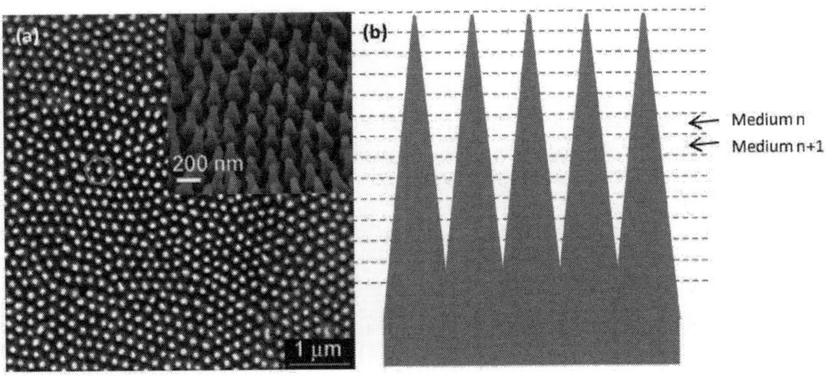

Figure 1. Antireflection structures. (a) Hexagonal arranged tapered pillars root in the surface of cicada wings in top view (tilted view of the pillar array is showed in the inset) [**19**]; (b) The tapered pillars lead to gradual refraction index variation in view of effective theory and then minimize the reflection according to Fresnel relations.

2.1.2. Bio-Inspired Fabrications

Inspired by nature, many efforts are made in exploring techniques for fabricating the anti-reflection nanostructures and a variety of methods, e.g., conformal evaporated-film-by-rotation (CEFR), colloidal lithography, self-masked dry etching, nanoimprint lithography (NIL), ALD and other approaches, are realized [16].

Oblique angle deposition (OAD) is usually employed to fabricate anisotropic film which is originated from the oblique growth of contained nanorods with a tilted angle to the substrate surface normal. The so-called CEFR method, which rotates the substrates at a high speed under OAD, leads to the straight growth of a dense array columns to substrate surface rather than helical structures which are formed under low rotation rate. That is, CEFR method is suitable for conformal replication of the photonic structure with a curved surface even under thick film deposition. With compound eyes of the fruit fly as bio-templates, it is reported artificial replica is successfully fabricated and hence similar optical features are inherited, respectively [22].

Many lithography techniques are applied for fabricating antireflection structures, among which colloidal lithography is amuch simpler approach [23]. With colloidal crystals asmasks, the silicon substrate is etched by reaction ion etching (RIE). During the fabrication duration, the colloidal spheres are etched by RIE gradually firstly, leading to a reduced transverse cross section of the spheres and thus an increasing exposure of the substrate. Attributed to the features of RIE, the etching rates of the apex and the junction parts of the spheres are not uniform, resulting in the etching morphology modification from frustum to cone arrays on the substrate finally [24].

Electron cyclotron resonance (ECR) plasma etching technique is employed by researchers to fabricate antireflection structures with much higher aspect ratio surfaces [25–27]. With the selected gas-mixture consisting of SiH_4, CH_4, Ar, and H_2, one step and self-masked dry etching are realized for fabricating high density nanotip arrays on a 6-inch silicon wafer. The fabrication progress is illustrated in the schematic diagram of Fig. 2(a). In brief, SiC clusters, which size and density can be tuned via process temperature, gas pressure, and composition, are formed on surfaces of the silicon substrate due to the reaction of SiH_4 and CH_4 plasma. Ar and H_2 are responsible for the dry etching process. The SiC clusters then act as nanomasks or nanocaps to protect the underlying substrate frometching, thus forming an aperiodic array of silicon nanotips with their lengths varying from ~1000 nm to ~ 16 μm finally (Fig. 2(b) and (c)). Even superior to natural prototypes, the biomimetic antireflection structures exhibit striking omnidirectional low reflection shown in Fig. 2(d) over a broad range of wavelengths from ultraviolet to terahertz region, irrespective of polarization.

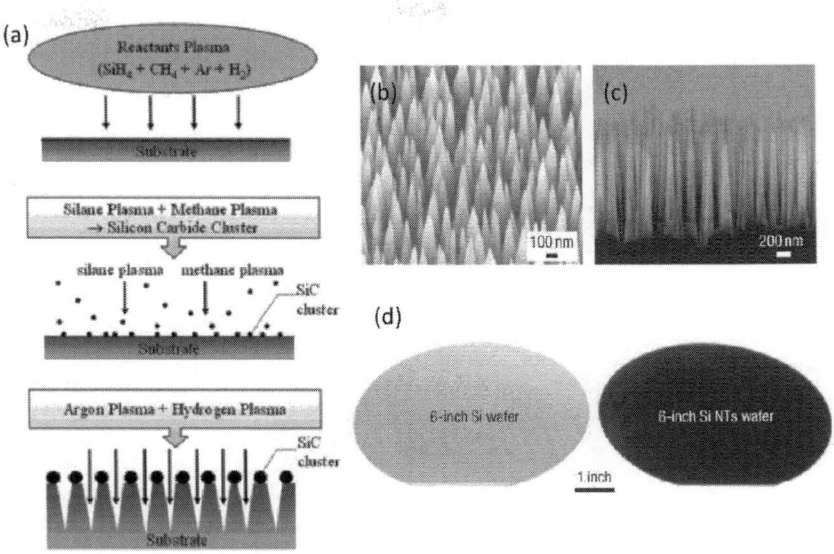

Figure 2. Improved antireflection of biomimetic nanotips by ECR plasma etching technique [**25**, **27**]. (a) Schematic diagram for the silicon nanotip formation on silicon wafer; (b) and (c) show a tilted top view and a cross-sectional view of silicon nanotips, respectively; (d) Compared with polished silicon wafer (left), 6-inch silicon wafer coated with silicon nanotips (right) show greatly improved broadband and quasi omnidirectional anti-reflection.

Avoiding time-consuming and complicated mask fabrication, scientists also attempt to directly use cicada wings or insect eyes as bio-templates [19, 28, 29]. For example, the nipple arrays on wing surfaces are stamped under certain pressure on glass-phase PMMA, which is at higher degree than its glass-transition temperature, supported by silicon wafer. A release process makes the polymer reverse nanostructures of the bio-templates. With the patterned PMMA as a mask or a mold, inverse or similar structures of the cicada wing are achieved by RIE or thermodeposition. A schematic diagram of the NIL, the fabricated structures, and the natural templates for comparison are illustrated in Fig. 3(a), Fig. 3(b), and Fig. 3(c), respectively. It is also worthy to note that an extra advantage of using bio-templates in NIL process is the notable low-surface-tension, which is vital for the release process, due to the wax layer commonly found on surfaces of plants and insects.

Taking advantages of accurate thickness control and three-dimensional (3D) fabrication of ALD, conformal replica is accomplished after ALD growth and sintering the hybrid structures with fly eyes as bio-templates, achieving similar anti-reflective features in the artificial analog finally [29].

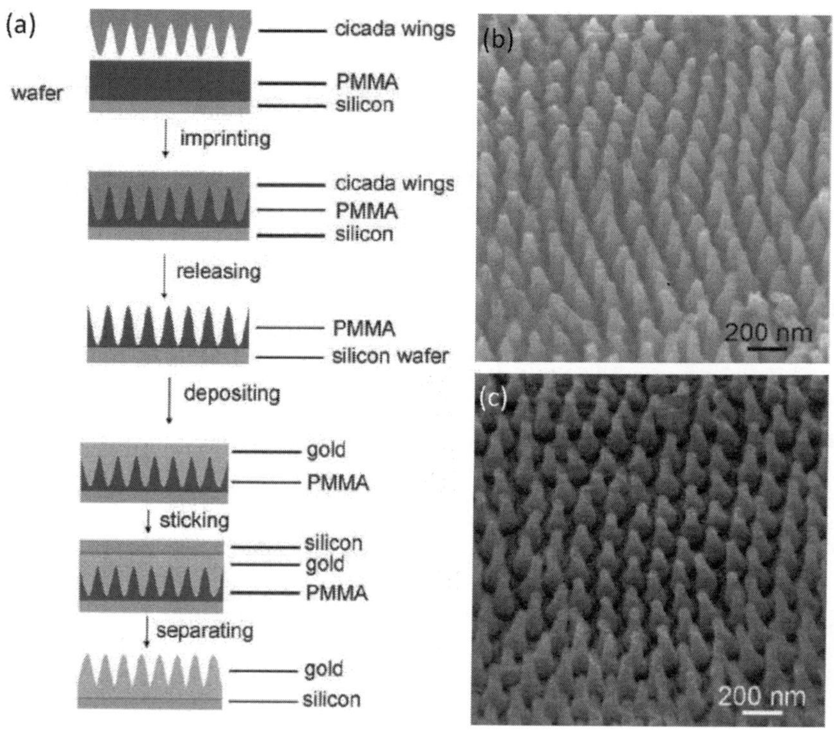

Figure 3. Cicada wing structure fabrication by NIL [**19**]. (a) Schematic diagram of NIL using cicada wings as bio-templates; (b) The fabricated structures and (c) the natural photonic structures show similar morphology from the SEM images.

2.1.3. Potential Applications

Due to the high reflectance of silicon solar cells (more than 30%) induced by the high index contrast of silicon and air according to the Fresnel equations, scientists are already aware of the vital roles of high-quality antireflection coatings at early ages of solar cell fabrications

[30]. Inspired by antireflection structures found in moth eyes, nanodomes or similar architectures are reproduced on surfaces of solar cells, leading to a dramatic light absorption increase and therefore a superior efficiency improvements than that of quarter-wavelength antireflection coating [31–33], as shown inFig. 4(a) -(d). The recent theoretical investigations even report a high light trapping close to the Yablonovitch limit in the silicon solar cell by optimizing a double-sided antireflection structure design (Fig. 4(e) and (f))[34].

The biomimetic antireflection structures can also play a key role in light extraction of light-emitting devices (LEDs) [35–37]. Because of the total internal reflection and the waveguiding modes in the glass substrate, only about 20% amount of the generated light can irradiate from the LEDs. By fabricating silica cone arrays on the surfaces of the ITO glass substrate to modulate the above two bottleneck factors, the light luminance efficiency of white LEDs is significantly improved by a factor of 1.4 in the normal direction and even larger enhancement for large viewing angles.

Another fascinating application of the inspired antireflection structures is the use in the micro Sun sensor for Mars rovers [38]. On the basis of the recorded image by an active pixel sensor, the location coordinates of the rover can be calculated. However, the ghost image originating from the multiple internal reflection of the optical system leads to severe limitation of the accuracy. By fabricating dense nanotip arrays on the surfaces of the sensor, the internal reflection is minimized to be nearly 3 orders of magnitude lower than that of no treatments, resulting in a more reliable three-axis attitude information.

2.2. Color-Tunable Devices

2.2.1. Prototypes

Besides the well known coloration change strategy via migrations and volumes change of pigment granules such as chameleons, nature develops a second approach which is known as structural coloration change (SCC). By varying photonic structure characterizations, incident light angle, or the refraction index contrast of the color-produced optical system via the environmental stimuli, reversible coloration changes, which are basically passive, are revealed in fishes, beetles, and birds [39–46]. Most structural basis of SCC are attributed to the one-dimensional (1D) reflectors. For example, the damselfish *Chrysiptera cyanea* can change its color from blue to ultraviolet rapidly under stressful conditions, which is triggered by the simultaneous change in the spacing of adjoining reflecting plates made of guanine in the iridophore cell [47–49]. In insect world, the coloration change of longhorn beetles *Tmesisternus isabellae* from golden to red is revealed to originate from the swollen multilayer after water absorption [50] (Fig. 5(a), (b) and (c)). Another origin of structural coloration change is the tilted angle variability of the nanoplates with respect to incident light, which is found in tropic fish neon tetra *Paracheirodon innesi* [51–53] (Fig. 5(d), (e), and (f)). Physically, the underlying mechanism of the coloration change in the mentioned 1D biological photonic structures can be understood according to the given formula

$$\lambda_{max} = 2(n_1 d_1 cos\theta_1 + n_2 d_2 cos\theta_2), \quad (3)$$

where d is the layer thickness, n is refraction index, and θ is angle of refraction. The subscripts represent the layer index. The angle of refraction at different layers θ_1 and θ_2 can be obtained from Snell's law $n_1 sin\theta_1 = n_1 sin\theta_1 = sin\theta_0$, where θ_0 is the incident angle from air. From the equation, it is easy to elucidate that either n, d, or θ_0 is related to the optical path in the multilayer and thus leads to shifting interference peaks at different wavelengths.

In biological system, high-dimensional photonic structures responsible for the coloration change are also discovered, though they are rare. An intriguing example is hercules beetles *Dynastes hercules*which can alter their appearance from khaki-green to black providing the ambience changes from dry to a high humidity level [54–56] (Fig. 5(g) and (h)). The 3D photonic crystal structures (nanoporous structures) are filled with water instead of air voids in dry status, rendering different refraction index contrast and thus the variation of Bragg scattering (Fig. 5(i), here only 2D cross section is illustrated). However, the underlying physics of coloration change induced by high-dimensional photonic structures is no other than that of their 1D counterparts.

2.2.2. Bio-Inspired Fabrications And Applications

Due to the obvious appearance changes which are easy to be picked up with the naked eye, color-tunable devices are explored to identify the status changes by temperature, vapor, solvent, humidity in ambience, the applied mechanical force, electric field, magnetic field and etc.

Besides the sensors, some novel writing system ('paper and ink') are developed. The key idea is to modify period (d), refraction index (n), view angle (θ) or their combinations of photonic structures, just like nature shows us.

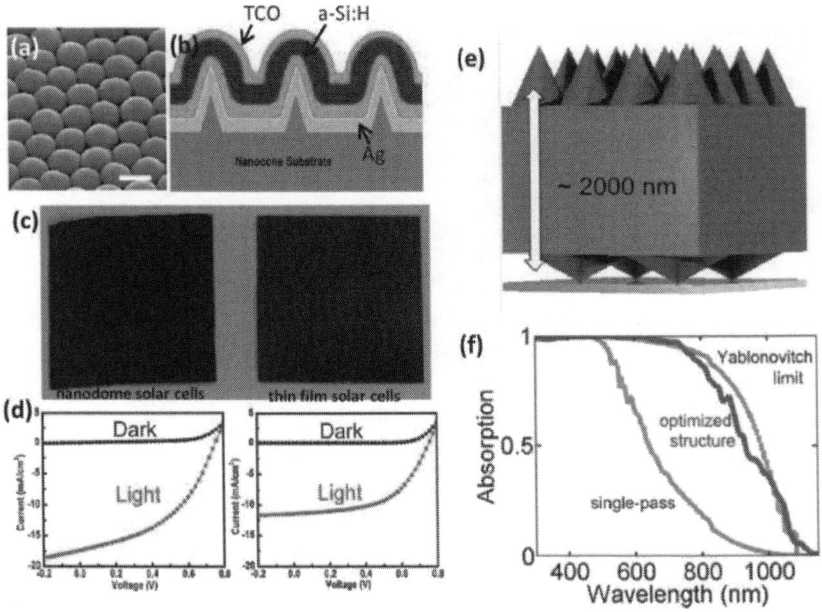

Figure 4. Color online) Solar cells coating antireflection structures [**32**, **34**]. (a) Nanodome solar cells in top view. Scale bar 500 nm; (b) Schematic diagram of the cross-sectional structures of the solar cells; (c) The photographs of nanodome solar cells (left) and flat film solar cells (right); (d) Dark and light I-V curve corresponding to (c); (e) and (f) The optimized double-sided nanostructure yields a photocurrent close to the Yablonovitch limit at an equivalent thickness of 2 μm.

2.2.2.1. *d variation*

Because of relatively simple control by the environmental stimuli and large spectral variation which can be recognized by the naked eye, approaches on the modulation of photonic structure period are always of scientist interests in obtaining tunable color applications.

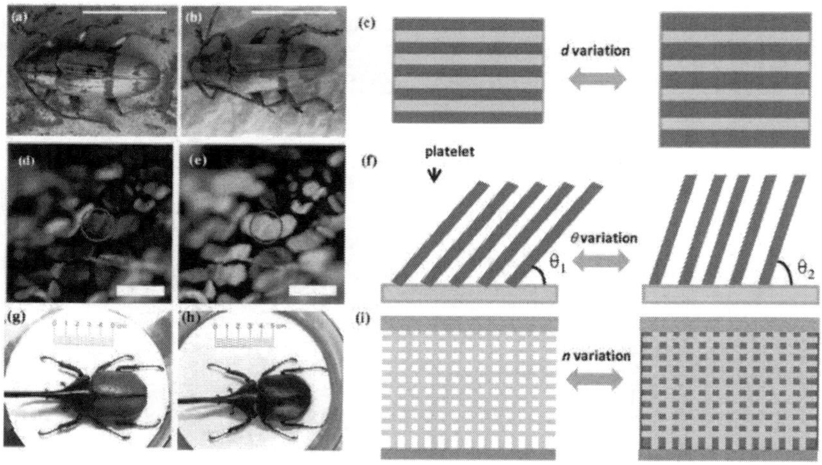

Figure 5. Color online) Color tuning mechanisms found in nature [**50**, **53**, **56**]. The coloration of (a) longhorn beetles, (d) iridophore of tropic fish neon tetra, and (g) hercules beetles can reversibly change their coloration to (b), (e), and (h), which are intrinsically induced by (c) period *d*, (f) tilting angle *θ*, (i) refraction index *n* variation, respectively. Scale bars: (a) and (b) 10 mm, (d) and (e) 20 μm.

'M-Ink' is a mixture of colloidal nanocrystal clusters (CNCs), solvent and photocurable resin (Fig. 6(a)). With the superparamagnetic Fe_3O_4 nanocrystals encapsulated by silica shell, 'M-Ink' can response to external magnetic fields. The role of the resin is to provide repulsive force which balances the attractive force of the CNCs. Without external magnetic fields, CNCs are randomly dispersed (infinite period) in liquid resin. The exhibited coloration is consistent with the magnetite, to be brown. After applying magnetic fields, the CNCs are assembled to form chain-like structures along the magnetic field lines (Fig. 6(b)). The additional magnetic force, the intrinsic force among the CNCs, and the repulsive force by resin establish dynamic balance with variation of the external magnetic fields, tuning the distance between the neighboring CNC (finite period). The switchable period then determines the color of the light diffracted from the CNC chain, leading to a full color show (Fig. 6(c)). The final step is to fix the desired coloration. After exposure to ultraviolet (UV) light at different exerted magnetic fields locations, the chain-like CNCs can be frozen in the solidified resin instantaneously, remaining the periods of the chains undistorted and accomplishing high-resolution color pattern fabrication (Fig. 6(d)) [57].

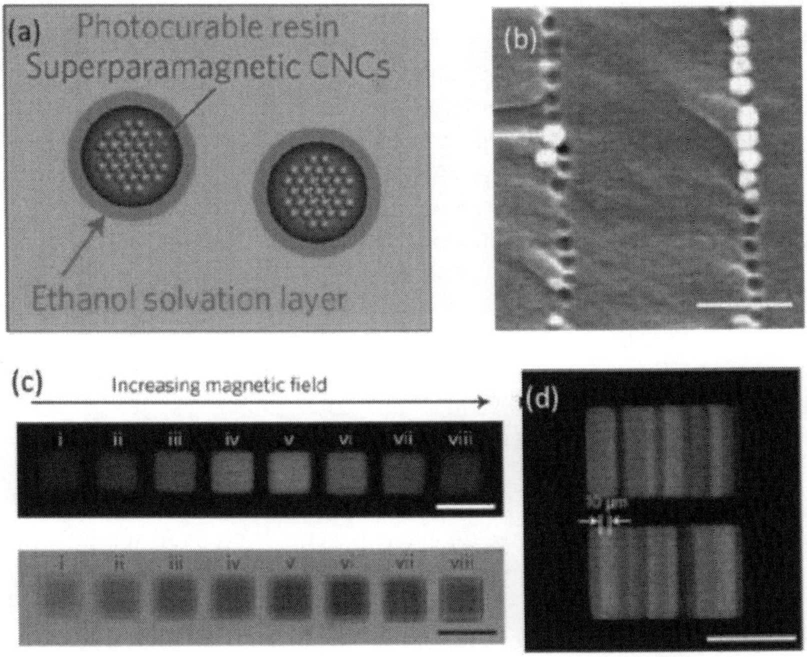

Figure 6. Color online) Structural color printing using 'M-Ink' [**57**]. (a) Three-phase material system of 'M-Ink'; (b) 'M-Ink' particles align in a chain under external magnetic field, which acts as basic coloration units of color pattern after fixing by UV light; (c) Reflection images of multicolored structural colors (upper) and transmission photographs of the same sample (below) by gradually increasing magnetic fields; (d) High-resolution multiple structural color patterns. Scale bars: (b) 1μm, (c) and (d) 100 μm.

With similar principle, the electric field-driven tunable color sensor is also realized by highly charged polystyrene (PS) colloids which form non-close-packed face-centered cubic (fcc) lattice [58]. Tuning the period along [111] direction by the balance of the exerted electrostatic force and the repulsive force, the exhibited coloration changes as a result of the applied electric field. The so-called 'P-Ink' is an electroactive material which consists of inverse opal inside polyferrocenylsilane (PFS) derivatives matrix. Such ink fabrication includes 3 primary steps: An opal film made of silica spheres is deposited onto glass substrate first by self-assembly; UV light is exposed to the sample in order to solidify the matrix and then form a stable PFS/silica composite; With diluted HF, inverse opal structure are realized in the elastomeric polymer matrix. By applying tunable voltage, macroscopic swelling and shrinking of the polymer matrix and microscopic Bravais lattice change responsible for the reverse coloration occur [59]. Stimulated by electrical forces, quite a few switchable coloration devices or sensors based on other materials or circumstances can be found elsewhere [60–62].

Many pressure-based photonic and even laser devices are reported [63–66]. Using monodispersed PS spheres to form cubic close packing (ccp) structures which are embedded in polydimethylsiloxane (PDMS) matrix, reverse colors are observed simply by stretching and releasing the rubber sheet. Upon mechanical stress, the lattice is elongated along the applied force direction, while the interplanar spacing in the perpendicular direction (i.e. distance between the (111) planes) decreases because of the nearly invariance volume of the rubber. The compressed distance leads to a blue-shift, e.g., from red to green [63]. Such opal rubber is believed to have practical applications such as a color indicator, tension meter or elongation strain sensor. The inverse opal structures (filled by air voids) in elastomer network are also synthesized [64, 67] (Fig. 7(a)). The porous elastomeric photonic crystals (EPCs) show highly reversible optical response to compressive force, e.g., 60 nm spectral blue-shift under a compressive pressure of ~ 15 kPa in the structures having 350-nm void size. Although the coloration change can be attributed to the Bravais lattice deformability, like the mechanism mentioned before. However, it is especially noteworthy that porous EPCs remain nearly undeformed in orthogonal directions when an external pressure is exerted in one direction, which can be ascribed to the high filling factor of air voids. The air voids enable the distortion of the cross-sectional void spaces from roughly circular to elliptical shape and a reduction of the air volume fraction under pressure. The elastic deformation feature of such structures helps to reduce the redistribution of stress along lateral directions when compressed by a patterned surface, leading to novel biometric applications such as the fingerprint recognition devices, as shown in Fig. 7(b). Additionally, air voids of porous EPCs provide a platform to incorporate with other functional materials for us to explore new applications. For instance, filling PbS quantum dots in the air voids, the photoluminescence (PL) emission, which leads to many potential applications in the near-infrared region, can be modified by overlapping with the forbidden bandgap of the inverse opal structures (Fig. 7(c)).

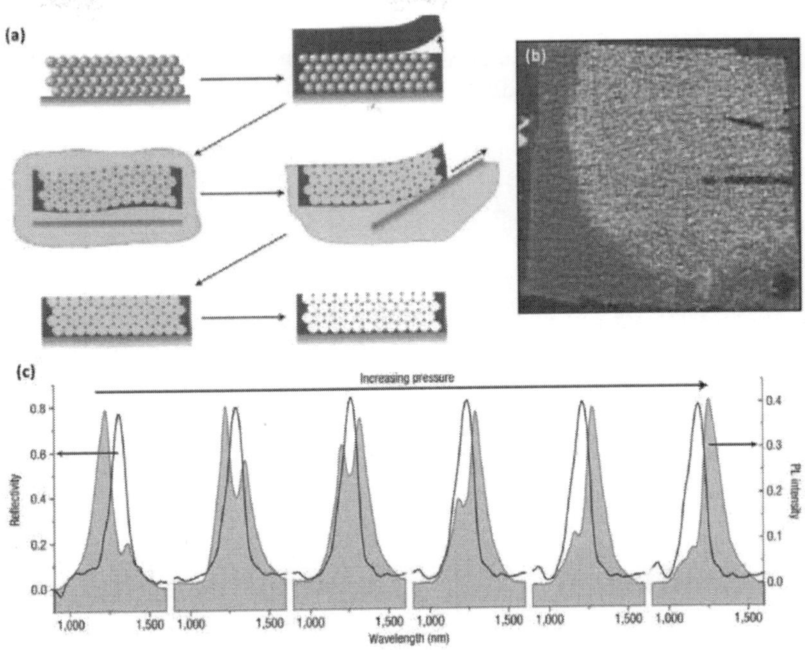

Figure 7. Color online) From color fingerprinting to the control of photoluminescence in EPC films [**67**]. (a) Schematic diagram of the elastic inverse opal structure fabrication; (b) A captured still image of the EPC film under compression by an index finger; (c) NIR-emitting PL emission of colloidal PbS quantum dots which are incorporated into voids of the EPC can be tuned by overlapping the forbidden gap (solid black line) with the PL peak (grey filled curve).

The chemical solvents are also used to be as stimuli. An interesting example is invention of new type 'photonic paper/ink' system [68, 69]. With novel soft materials consisting of closely packed PS and polymer elastomer (colloids and PDMS), the exhibited coloration can be altered reversely by immersing the materials into silicone liquid ('writing process') and an evaporation process ('erasing process'). The spacing between the (111) planes is adjusted by the strength of interaction between PDMS matrix and the contained silicone oligomers with different molecule weight in the liquid. With different solvents ('ink'), the swelling and shrinkage of the matrix showa featured reversible shift of Bragg diffraction peak. Amultilayer based on alternating Teflon-like layer and Au nanoparticle/Teflon-like layer composite layer operates not in visible but optical telecommunication wavelength range [70]. When the structure is exposed to different organic solvent vapor (e.g.

acetone, ethanol, methanol, water, chloroform, and etc.), the molecules enter the metal/polymer composite inside the holes and microvoids in its structure, resulting in the swelling up, e.g., for acetone vapors at a molar fraction of 0.25, with an relative increase of 12.5% in thickness to an equilibrium state while leaving the Teflon-like layer unchanged due to its inert chemical features. The measured reflectance show a large variation of 0.2 μm, which is advantageous to the detection of the organic/inorganic vapors.

Other external stimuli such as UV light, heat, or chemical reaction are applied to trigger and fabricate coloration sensitive materials as well by reversely controlling the spacing of the responsible photonic structures. Detailed information can be found in some references and recent reviews [14, 15, 61,71–74].

2.2.2.2. n variation

Besides *d*, the refraction index *n* is another attribute of photonic structures. Different approaches to alter refraction index, in which infiltration is most regular, are revealed in order to achieve novel visual applications. Inspired by beetles *D. hercules*, 3D nanoporous structures are fabricated by so-called dip-coating deposition, achieving humidity sensing [75]. The 3D architecture, which is inverse opal structure, is reported to have its reflective peak red-shifting 14 nm when the relative humidity changes from 25% to 98%. It is worth noting that after treatments by O_2 plasma, the 3D photonic structures exhibit large bandgap shift of 137 nm and dramatic coloration change from bluish green in dry state to red in fully wet state. The modified hydrophilic feature play an important role in water collection in air voids at different level of humidity, leading to larger variation of refraction index contrast. The other clew for the remarkable coloration change can be ascribed to the high filling factor of air voids of inverse opal structures. The high value of ~ 76% results in a wider modulation of refraction index contrast by the amount of solvent absorption. Tuning hydrophilic features by chemical groups, the inverse opal film can even be used as 'Watermark-Ink' system [76]. In the studies, the inside porous structures are functionalized with chemical group (*R*1, *R*2, *R*3, or *R*4) through vaporizing an alkylchlorosilane. The chemical functionalities of the structures then are erased and the surface is reactivated by O_2 plasma exposure, leaving a patterned functionalized region where is masked by a PDMS polymer. With iterations of such kind of functionalization and reactivation, the opal structures are locally patterned by different chemical groups, leading to differentiate hydrophilic feature (Fig. 8(a) and (b)). When immersing the film into specific fluids, regulated infiltration by the wettability occurs spatially in the film and thus surveys different optical responses (i.e. different patterns, Fig. 8(c) and (d)). The fabricated structures are expected to have applications in encryption as well as colorimeter.

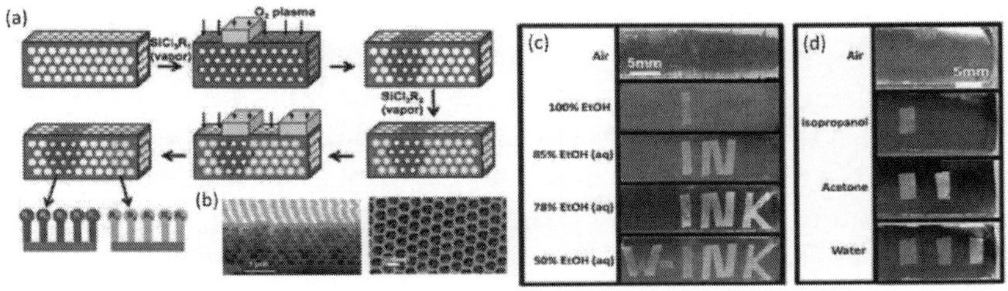

Figure 8. Color online) Silica inverse opal films of 'Watermark-Ink' system [**76**]. (a) Schematic procedure of chemical encoding; (b) SEM images of the fabricated inverse opal structures in cross-sectional view (left) and top view (right); (c) Optical images of the fabricated film in which the word 'W-INK' is encoded via functional chemical groups on surfaces. (d) Optical images of different encoded patterns under different solvents.

Besides infiltration, phase-transition materials can also induce refraction index variation and switchable coloration, providing the transition conditions are satisfied. The best-known phase-transition material maybe is liquid crystals (LCs). Above the phase-transition temperature of 34°, LCs change their nematic phase which is anisotropic to isotropic phase, leading to a significant change of refraction index and coloration, e.g., in inverse opal structures. By mixing the active material into LCs, sensitive reflection and polarization triggered by UV light are reported by a series of subsequent researches [77–79]. Besides LCs, various sensitive materials, including Ag_2Se, WO_3 and ferroelectric ceramics, are also found functionally in incorporating with diversified photonic structures to obtain tunable structural coloration materials and thus fabricate thermo- or electro-sensors [80–83].

2.2.2.3. θ variation

Iridescence is a characteristic of structural coloration, which is determined by band dispersions of photonic structures. With various incident angle or observation direction, the perceived coloration is changing. Hence, it is feasible to explore novel tunable color devices by tuning orientations of the photonic structures [84–86]. Just like the fish neon tetra *P. innesi*, magnetochromaticmicrospheres alter their appearance by tilting the inner photonic structures with respect to the incidence (Fig. 9). As mentioned before, with superparamagnetic Fe_3O_4 particles coated by silica in PEGDA emulsions, the applied magnetic field guides to form photonic chains, inwhich the interpaticle distances depend on the balance of the attractive and repulsive force. Polymerized by UV light, microspheres are solidified and the inner periodic photonic chains, which give rise to structural coloration, are fixed permanently. Dispersing the microspheres into liquids, various colors are observed in top view due to the random orientations of the photonic chains in the microspheres. By applying external magnetic fields, the microspheres tend to rotate in order to keep the magnetic chains inside in the direction of the magnetic vector, leading to a homogenous coloration of green exhibited. The feature of the switchable coloration 'on' and 'off' status by the external stimuli is believed to have applications in color display, signage, bio- and chemical detection, and magnetic field sensing.

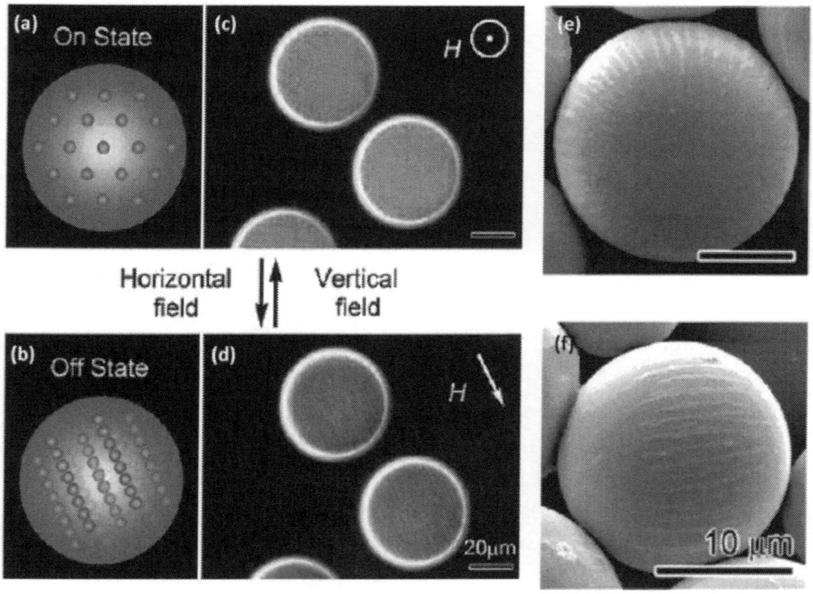

Figure 9. Color online) Magnetochromatic microspheres switched between 'on' and 'off' states by rotating external fields [**84**, **85**]. Schematic illustrations, optical images, and the corresponding SEM photographs of 'on' and 'off' states are (a), (c), (e), and (b), (d), (f), respectively.

2.3. Structural Color Mixing And Applications

2.3.1. Prototypes

Structural coloration results fromthe interaction of light and photonic structureswith featured size of visible wavelengths. It is even more widespread than pigmentary coloration in animal world. Some literatures have well reviewed the field comprehensively. In the Chapter, we do not plan to pay attention to the overall structural coloration but only focus on a specific subject 'structural color mixing' [87–92]. In tiger beetles *Cicindela oregona*(Fig. 10(a)-(c)) [87, 88], the honeycomb-like pits are found on surfaces of the elytra. Under microscope, the brown morph actually includes blue-green patches in the red background, while the black one consists of magenta patterns surrounded by dull green. The microscopic colors are the results of light interference by the multilayer structures in the cuticle. However, due to the colored patches (40-80 μm across) are too small to be resolved by the unaided eye (the resolution d is determined by the Rayleigh criterion $d=l\times0.61\lambda/D$, where l is distance of distinct vision, λ the wavelength and D the average pupil diameter of humankind), the perceived coloration is a mixture of the discrete colors, leading to a totally different exhibition fromthat equippedwith microscope. From the investigations of photonic structures of beetles *Chlorophila obscuripennis* (Fig. 10(d)-(f)), we revealed the color is a juxtaposition in a smaller region (~ $10\times10\mu m^2$) by green on the ridges and cyan in centers of the pits. Furthermore, the pits on the elytra surface give rise to diffused light reflected over a wide large of angles, leading to inconspicuous coloration shown [91]. In butterflies *Papilio palinurus* (Fig. 10(g)-(i)), the scanning electronic micrographs show surfaces in the scales comprise a similar two-dimensional (2D) pits (4-6 μm in diameter and 3 μm at the greatest depth). The flat regions between and in pits appear yellow, and the inclined region contributes to blue color. Because of the limitation of human eyes' resolution, the butterfly displays a mixture color of green, i.e. yellow plus blue turns to be green. Due to sufficient depth of the pits, light which normally incidents on the inclined side can experience dual reflection and is back-reflected. The retro-reflection is found to play a key role in the polarization conversion of incident light, which is crucial in some novel optical applications. Thanks to symmetric feature of the pits, no macroscopic polarization effects can be observed [90]. On the cover scale surfaces of butterflies *Suneve coronata*, however, the natural occurring triangular grooves array not symmetric but in 1D way with a period of ~ 2 μm, giving rise to polarization conversion of the normal incident light in a specific direction. Intriguingly, it is revealed the coloration macroscopically remains unconscious change under different polarized illumination. The reason is the curly scales, which induce polarization conversion between the neighboring rows of scales but in the orthogonal direction. The nanostructures having different magnitude in size therefore cancel out the macroscopic polarization effects (Fig. 12(a)-(d)) [89].

Figure 10. Color online) Structural Color Mixing in Nature [**87**, **90**, **91**]. Photographs, optical microscopy and SEM images of tiger beetles *C. oregona* ((a), (b), (c)), long-jointed beetles *C. obscuripennis* ((d), (e), (f)), and swallowtail butterflies *P. palinurus* ((g), (h), (i)), respectively.

2.3.2. Bio-Inspired Fabrications And Applications

Structural coloration may be especially crucial for future color and related industry because of the non-fading feature (if the photonic structures are undeformed) [93, 94] and environmental friendliness. Naturally, structural coloration mixing inherits the advantages. Mimicking the nature, mixed structural color and its application can be obtained. For example, PS colloids with a diameter of 5 μmare assembled on a gold-coated silicon substrate. A 2.5-μm-thick layer of platinum or gold is then deposited to fill the interspaces of the colloids by electrochemical approach, creating a negative replica. After removal of the PS colloids by ultrasonic waves and a sputtering thin carbon film, a multilayer of quarter-wave titania and alumina films is grown by ALD (Fig. 11(a)), inheriting the morphology of hexagonally arranged pits of the negative replica (Fig. 11(b)). The sample exhibits similar color mixing with that of butterflies *P. blumei* (Fig. 11(c) and (d)). Moreover, by modifying the surface morphology of the imitation (Fig. 11(e)), even visual information, e.g., a picture, can be encoded into the photonic structures which display a striking appearance change from pale blue in the specular direction to red in retro-reflection (Fig. 11(f) and (g)). The bio-inspired work is expected to find applications in security labelling field or color industry such as painting and coating [95]. Inspired by color mixing researches, some novel applications based on polarization conversion are designed [89]. For instance, by etching periodic triangular-like grooves on surfaces of a flat multilayer, the film displays a coloration of green, which is actually a mixed color from yellow in the flat region and blue in the grooves in normal incidence, as shown in Fig. 12(e). Because of the broken symmetry of the 1D structures, the blue color can be suppressed by using a polarizer on light path, leading to the exhibited coloration change to yellow (i.e. the hue in flat regions). That is, the grooved areas turn the light polarization $\pi/2$ and thus are shadowed by the inserted polarizer. Following the ideas, an encoded pattern which comprises grooved areas can be distinguished with the background consisting of perpendicular grooves via the luminosity level under different polarized light incidence (Fig. 12(f)). The two imaged examples inspired by the grooved structures of the butterfly show us great prospective in fields such as anti-fake and encryption.

2.4. Other Bio-Inspired Fabrications And Applications

Besides the main categories, some other bio-inspired photonic applications are also reported [96–100]. For examples, the natural scales of *Morpho sulkowskyi* are extremely sensitive to the different environmental vapors, which lead to dramatically improved responses as units of potential sensor applications compared with that of current devices[101]. Using black scales of the butterfly *Papilio paris* and *Thaumantis diores* as templates, hierarchically periodic microstructure titania replica was systhesized by chemical procedures. The high surface area inherited from the natural templates is great advantages to the light harvesting efficiency and dye sorption when the replica is used as photoanode in dye-sensitized solar cell [100]. With *Morpho* as bio-templates, alumina replicas show the potential applications in waveguides and beam-splitters under thin film coating by ALD and sintering [97]. Under thick film coating, the hybrid structure is found not to inherit the 'Christmas-tree' structures but develop a 'Pyramid-like' structures which have potential applications in light trapping [99]. With the complex photonic structures (multilayer plus 2D amorphous structures) of beetles *Trigonophorus rothschildi varians* as 'blueprints', the artificial counterpart which is fabricated by FIB etching holes through a 5×SiO/SiGe multilayer structure show similar optical features like non-specularity and only slightly angle-dependent reflectance [98]. Further bio-inspired work is undergoing which is stimulated by a recent revealed 3D architecture [102], aiming at the ultra-negative angular dispersion of diffraction and potential novel dispersive optical elements.

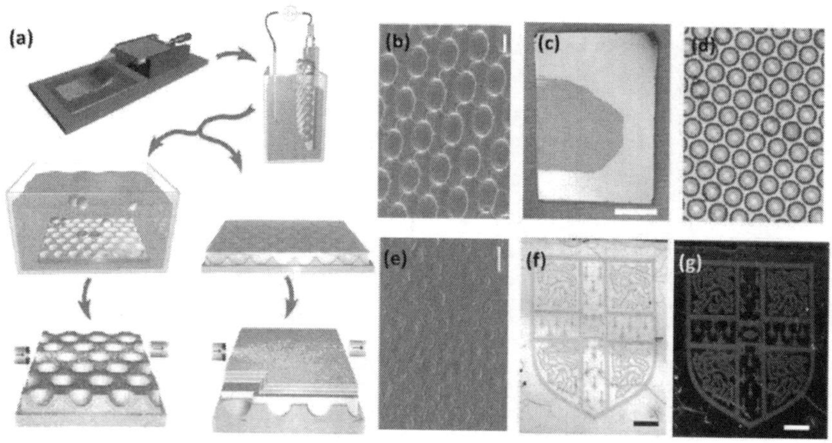

Figure 11. Color online) Mixed structural coloration and applications inspired by nature [**95**]. (a) Schematic diagram of artificial samples mimicking the scale structures of *P. blumei*; (b) SEM photographs in top view show the concavities on surface of the replica which is (c) green macroscopically but (d) resolved yellow and green microscopically; (e) Modifications of the concavities morphology by melting colloidal spheres embedded in the concavities lead to a striking change in color of the sample from blue to red viewed (f) in direct specular reflection and (g) in retro-reflection. Scales: (a) 2 μm, (c)(f)(g) 5 mm, and (e) 5μm.

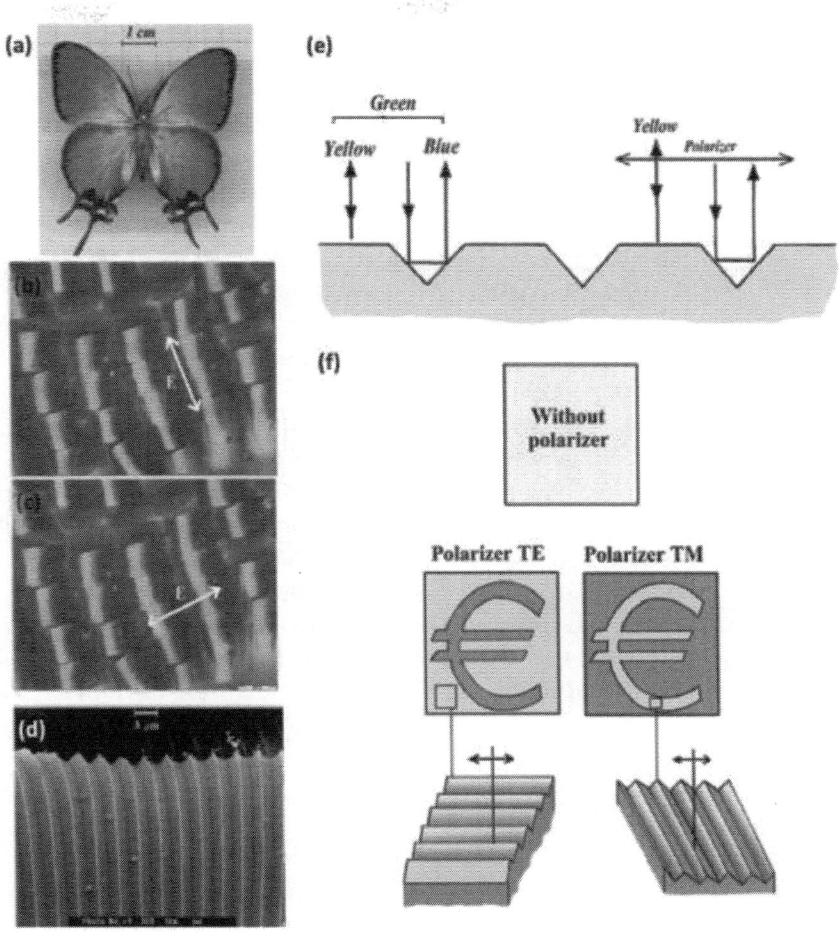

Figure 12. Color online) (a) Optical image of a male *S. coronata* butterfly in dorsal view; The appearances under (b) TM and (c) TE polarization light show distinguished difference of the blue component, which can be attributed to (d) the broken symmetry of the triangular grooves in the surface plane; Two kinds of anti-fake or encryption designs, which apply (e) colored effect or (f) grey level change linked to the polarization and the broken surface symmetry, are inspired [**89**].

3. PERSPECTIVES

In the Chapter, several important kinds of bio-inspired photonic applications are reviewed, including antireflection devices, color-tunable sensors, structural color mixing applications and etc. The nature nourishes scientists the functional optical applications either the blueprints of photonic architecture or directly the bio-templates. Due to the higher index of inorganic materials used, the mimicking photonic structures even show better optical performances as well as enhanced mechanical properties of high temperature tolerance, stability and infrangibility. The biomimetic applications are anticipated to help our life better in the near future. However, complicated photonic structures (e.g. those of high-dimensional, hierarchic, amorphous features in nature) still remains hardly reproduced or, if they are fabricated successfully, the efforts involved are so great using the traditional fabrication ways that optical devices can not commercially explored. Thorough physical mechanism understanding as well as better fabrication approach explorations may help to simply the structure fabrications, achieve similar optical functions and realize commercial applications. In addition, adding substances such as functional chemical groups, fluorescence particles, metal, or other active materials, the mimicking photonic structures allow the properties of interest to be augmented, which may open a new window of novel optical device exploration. Although the photonic biomimicry is in its infancy, we believe that the bio-inspired optical device would surely have profound impacts on our modern society.

REFERENCES

1. E. Yablonovitch, 1987Inhibited Spontaneous Emission in Solid-State Physics and ElectronicsPhys. Rev. Lett., 58

2. S. John, 1987Strong localization of photons in certain disordered dielectric superlatticesPhys. Rev. Lett., 58

3. J. D. Joannopoulos, S. G. Johnson, J. N. Winn, R. D. Meade, 2008Photonic Crystals: Molding the Flow of Light,nd Ed. Princeton University Press, Princeton and Oxford, USA.

4. D. L. Fox, (1976). Animal Biochromes and Structural Colours, University of California Press, Berkeley, USA.

5. M. Srinivasarao, 1999Nano-Optics in the biological world: beetles, butterflies, birds, and mothsChem. Rev., 991935 EOF

6. P. Vukusic, J. R. Sambles, 2003Photonic structures in biology.Nature424852 EOF5 EOF

7. A. R. Parker, 2005A geological history of reflecting opticsJ. R. Soc. Interface, 2

8. A. Seago, P. Brady, J. Vigneron-P, T. D. Schultz, 2009Gold bugs and beyond: a review of iridescence and structural colour mechanisms in beetles (Coleoptera)J. R. Soc. Interface, 6S165-S184.

9. M. D. Shawkey, N. I. Morehouse, P. Vukusic, 2009A protean palette: colourmaterials and mixing in birds and butterflies. J. R. Soc. Interface, 6S221-S231.

10. S. Kinoshita, S. Yoshioka, 2005Structural colors in nature: the role of regularity and irregularity in the structure.ChemPhysChem, 61442 EOF59 EOF

11. A. R. Parker, H. E. Townley, 2007Biomimetics of photonic nanostructures.Nat. Nanotech., 2347 EOF53 EOF

12. D. P. Pulsifer, A. Lakhtakia, 2011Background and survey of bioreplication techniques.Bioinsp. Biomim., 031001

13. W. Chung-J, J. Oh-W, K. Kwak, B. Y. Lee, J. Meyer, E. Wang, A. Hexemer, S. Lee-W, 2011Biomimetic self-templating supramolecular structures.Nature478364 EOF8 EOF

14. S. Fudouzi, 2011Tunable structural color in organisms and photonic materials for design of bioinspiredmaterials. Sci. Technol. Adv. Mater., 064704

15. Y. Zhao, Z. Xie, H. Gu, C. Zhu, Z. Gu, 2012Bio-inspired variable structural color materialsChem. Soc. Rev., 413297 EOF3317 EOF

16. Li. Y. (b, J. Zhang, B. Yang, 2010Antireflective surfaces based on biomimetic nanopillared arraysNano Today5117 EOF127 EOF

17. M. F. Land, D. E. Nilsson, 2001Animal EyesOxford University Press, Oxford, UK.

18. A. Yoshida, M. Motoyama, A. Kosaku, K. Miyamoto, 1997Antireflective nanoproturberance array in the transparent wing of a hawkmoth Cephanodes hylas. Zool. Sci., 14

19. G. Zhang, J. Zhang, G. Xie, Z. Liu, H. Shao, 2006Cicada wings: a stamp fromnature for nanoimprint lithography. Small, 12

20. M. Born, E. Wolf, (1999). Principles of Optics, 7th Ed. Cambridge University Press, Cambridge, UK.

21. J. Xi, Q. Schubert, M. F. Kim, J. H. Schubert, E. F. Chen, M. Lin, S. Liu, W. , J. A. Smart, 2007Optical thin-film materials with low refractive index for broadband elimination of Fresnel reflection. Nat. Photon., 1

22. R. J. Martín-Palma, C. G. Pantano, A. Lakhtakia, 2008Replication of fly eyes by the conformal-evaporated-film-by-rotation technique. Nanotechnology355704

23. S. Yang-M, 2006Nanomachining by colloidal lithography.Small, 4458 EOF75 EOF

24. S. J. Pearton, D. P. Norton, 2005Dry etching of electronic oxides, polymers, and semiconductors. Plasma Process. Polym., 2

25. C. Hsu, H. Lo, C. Chen, C. T. Wu, J. Hwang, D. Das, J. Tsai, L. Chen, K. Chen, 2004Generally applicable self-masked dry etching technique for nanotip arrayfabrication. Nano Lett., 3

26. C. Hsu, Y. F. Huang, L. C. Chen, S. Chattopadhyay, K. H. Chen, H. C. Lo, C. F. Chen, 2006Morphology control of silicon nanotips fabricated by electron cyclotron resonanceplasma etching. J. Vac. Sci. Technol. B, 1

27. Y. Huang, Chattopadhyay. S. Huang, Y. Jen, C. Peng, T. Liu, Y. Hsu, C. Pan, H. Lo, C. Hsu, Y. Chang, C. Lee, K. Chen, L. Chen, 2007Improved broadband and quasi-omnidirectional anti-reflection propertieswith biomimetic silicon nanostructures. Nat. Nanotech., 2

28. H. Gao, Z. Liu, J. Zhang, G. Zhang, G. Xie, 2007Precise replication of antireflective nanostructures from biotemplates. Appl. Phys. Lett., 12

29. J. Huang, X. Wang, Z. L. Wang, 2008Bio-inspired fabrication of antireflection nanostructures by replicating fly eyes. Nanotechnology025602

30. M. A. Green, 1987Higher Efficiency Silicon Solar Cells, Trans Tech Pub, Aedermannsdorf, Switzerland.

31. J. Zhao, M. A. Green, 1991Optimized antireflection coatings for high-efficiency silicon solar cellsIEEE Trans. Electron Devices, 81925 EOF1934 EOF

32. J. Zhu, C. Hsu, Z. Yu, S. Fan, Y. Cui, 2010Nanodome Solar Cells with Efficient Light Management and Self-CleaningNano Lett., 61979 EOF1984 EOF

33. H. L. Chen, S. Y. Chuang, C. H. Lin, Y. H. Lin, 2007Using colloidal lithography to fabricate and optimize sub-wavelength pyramidal and honeycomb structures in solar cells.Opt. Express, 1514793 EOF803 EOF

34. K. X. Wang, Z. Yu, V. Liu, Y. Cui, S. Fan, 2012Absorption enhancement in ultrathin crystalline silicon solar cells with antireflection and light-trapping nanocone gratingsNano Lett., 31616 EOF1619 EOF

35. M. Ishimori, Y. Kanamori, M. Sasaki, K. Hane, G. Xie, 2002Subwavelength antireflection gratings for light emitting diodes and photodiodes fabricated by fast atom beam etchingJpn. J. Appl. Phys., 414346 EOF4349 EOF

36. Li. Y. (a, F. Li, J. Zhang, C. Wang, S. Zhu, H. Yu, Z. Wang, B. Yang, 2010Improved light extraction efficiency of white organic light-emitting devices by biomimetic antireflective surfacesAppl. Phys. Lett., 153305

37. Y. M. Song, E. S. Choi, G. C. Park, C. Y. Park, S. J. Jang, Y. T. Lee, 2010Disordered antireflective nanostructures on GaN-based light-emitting diodes using Ag nanoparticles for improved light extraction efficiencyAppl. Phys. Lett., 093110

38. C. Lee, S. Y. Bae, S. Mobasser, H. Manohara, 2005A novel silicon nanotips antireflection surface for the micro Sun sensor.Nano Lett., 122438 EOF42 EOF

39. H. Cong, B. Yu, X. S. Zhao, 2011Imitation of variable structural color in paracheirodon innesi using colloidal crystal films.Opt. Express, 1312799 EOF808 EOF

40. N. F. Hadley, 1979Wax secretion and color phases of the desert Tenebrionid beetle Cryptoglossa verrucosa (LeConte).Science, 203367 EOF9 EOF

41. L. M. Mäthger, M. F. Land, U. E. . Siebeck, N. J. Marshall, 2003Rapid colour changes in multilayer reflecting stripes in the paradise whiptail, Pentapodus paradiseusJ. Exp. Biol., 206

42. E. Mc Clain, M. K. Seely, N. F. Hadley, V. Fray, 1985Wax blooms in Tenebrionid beetles of the Namib desert: correlations with environmentEcology66112 EOF

43. P. Jolivet, 1994Phsiological colour changes in tortoise beetles, In: Novel Aspect of the Biology of Chrysomelidae, Cox, M.L. & Petitpierre, E., (Eds.), page numbers (331-335), Kluwer Academic, Netherland.

44. J. P. Vigneron, J. M. Pasteels, D. M. Windsor, Z. Vertésy, M. Rassart, T. Seldrum, J. Dumont, O. Deparis, V. Lousse, L. P. Biró, D. . Ertz, V. Welch, 2007Switchable reflector in the Panamanian tortoise beetle Charidotella egregia (Chrysomelidae: Cassidinae)Phys. Rev. E., 031907

45. C. W. Mason, 1929Transient color changes in the tortoise beetles (Coleoptera: Chrysomelidae). Entomol. News, 45

46. C. M. Eliason, M. D. Shawkey, 2010Rapid, reversible response of iridescent feather color to ambient humidity. Opt. Express, 20

47. H. Kasukawa, N. Oshima, R. Fujii, 831 EOF7 EOF

48. H. Kasukawa, N. Oshima, R. Fujii, 1987Mechanism of light reflection in blue damselfish motile iridophore. Zool. Sci., 4

49. N. Oshima, R. Fujii, (1987, 1987Mobilemechanisms of blue damselfish (Chrysiptera cyanea) iridophores Cell Motil. Cytoskel., 8

50. F. Liu, B. Q. Bong, X. H. Liu, Y. M. Zheng, J. Zi, 2009Structural color change in longhorn beetles Tmesisternus isabellae.Opt. Express, 1816183 EOF91 EOF

51. J. N. Lythgoe, J. Shand, 1982Changes in spectral reflexions from the iridophores of the neon tetra.J. Physiol., 325

52. N. Oshima, 2005Light reflection in motile iridophores of Fish, In: Structural Colors in Biological Systemsa‚łPrinciples and Applications, Kinoshita, S. & Yoshioka, S., (Eds.), page numbers (211), Osaka University Press, Japan.

53. S. Yoshioka, B. Matsuhana, S. Tanaka, Y. Inouye, N. Oshima, S. Kinoshita, 2011Mechanism of variable structural colour in the neon tetra: quantitative evaluation of the Venetian blind modelJ. R. Soc. Interface, 856 EOF66 EOF

54. H. E. Hinton, G. M. Jarman, 1972Physiological colour change in the Hercules beetleNature238160 EOF161 EOF

55. H. E. Hinton, G. M. Jarman, 1973Physiological colour changes in the elytra of the Hercules beetles, Dynastes hercules. J. Insect Physiol., 19533 EOF549 EOF

56. M. Rassart, J. Colomer-F, T. Tabarrant, J. P. Vigneron, 2008Diffractive hygrochromic effect in the cuticle of the hercules beetle Dynastes herculesNew J. Phys., 033014

57. H. Kim, J. Ge, J. Kim, S. Choi, H. Lee, H. Lee, W. Park, Y. Yin, S. Kwon, 2009Structural colour printing using a magnetically tunable and lithographically fixable photonic crystalNat. photon., 3534 EOF540 EOF

58. T. S. Shim, S. Kim-H, J. Y. Sim, J. Lim-M, S. Yang-M, 2010Dynamic Modulation of Photonic Bandgaps in Crystalline Colloidal Arrays Under Electric FieldAdv. Mater., 224494 EOF4498 EOF

59. D. P. Puzzo, A. C. Arsenault, I. Manners, G. A. Ozin, 2009Electroactive Inverse Opal: A Single Material for All Colors.Angew. Chem., Int. Ed., 47943 EOF7 EOF

60. A. C. Arsenault, D. P. Puzzo, I. Manners, G. A. Ozin, 2007Photonic-crystal full-colour displays. Nat. Photon., 1

61. J. J. Walish, Y. Kang, A. Mickiewicz, E. L. Thomas, 2009Bioinspired electrochemically tunable block copolymer full color pixelsAdv. Mater., 213078 EOF3081 EOF

62. K. Hwang, D. Kwak, C. Kang, D. Kim, Y. Ahn, Y. Kang, 2011Electrically tunable hysteretic photonic gels for nonvolatile display pixelsAngew. Chem., Int. Ed., 506311 EOF6314 EOF

63. H. Fudouzi, T. Sawada, 2006Photonic rubber sheets with tunable color by elastic defornmation. Langmuir, 3

64. K. Sumioka, H. Kayashima, T. Tsutsui, 2002Tuning the Optical Properties of Inverse Opal Photonic Crystals by DeformationAdv. Mater., 181284 EOF1286 EOF

65. S. Fudouzi, T. Kanai, T. Sawada, (2011, 2011Widely tunable lasing in a colloidal crystal gel film permanently stabilized by an ionic liquid. Adv. Mater., 33

66. A. C. Arsenault, V. Kitaev, I. Manners, G. A. Ozin, A. Mihi, H. Míguez, 2005Vapor swellable colloidal photonic crystals with pressure tunabilityJ. Mater. Chem., 15133 EOF138 EOF

67. A. C. Arsenault, T. J. Clark, G. V. Freymann, L. Cademartiri, R. Sapienza, J. Bertolotti, E. Vekris, S. Wong, V. Kitaev, I. Manners, R. Z. Wang, S. John, D. Wiersma, G. A. Ozin, 2006From colour fingerprinting to the control of photoluminescence in elastic photonic crystalsNat. Mater., 5179 EOF184 EOF

68. Fudouzi. H. (a, Y. Xia, 2003Photonic papers and inks: color writing with colorless materialsAdv. Mater., 11892 EOF

69. Fudouzi. H. (b, Y. Xia, 2003Colloidal crystals with tunable colors and their use as photonic papersLangmuir, 239653 EOF

70. A. Convertino, A. Capobianchi, A. Valentini, N. M. C. Emilio, 2003A new approach to organic solvent detection: high-reflectivity Bragg reflectors based on a gold nanoparticle/Teflon-like composite materialAdv. Mater., 131103 EOF

71. J. M. Weissman, H. B. Sunkara, A. S. Tse, S. A. Asher, 1996Thermally Switchable Periodicities and Diffraction from Mesoscopically Ordered MaterialsScience5289959 EOF

72. J. H. Holtz, S. A. Asher, 1997Polymerized colloidal crystal hydrogel films as intelligent chemical sensing materials.Nature389829 EOF32 EOF

73. Z. Hu, X. Lu, J. Gao, 2001Hydrogel opals. Adv. Mater., 22

74. B. Gates, S. H. Park, Y. Xia, 2000Tuning the photonic bandgap properties of crystalline arrays of polystyrene beads by annealing at elevated temperaturesAdv. Mater., 9653 EOF656 EOF

75. J. H. Kim, J. H. Moon, S. Lee-Y, J. Park, 2010Biologically inspired humidity sensor based on three-dimensional photonic crystalsAppl. Phys. Lett.,103701

76. I. B. Burgess, L. Mishchenko, B. D. Hatton, M. Kolle, M. Loncar, J. Aizenberg, 2011Encoding complex wettability patterns in chemically functionalized 3D photonic crystalsJ. Am. Chem. Soc., 13312430 EOF12432 EOF

77. S. Kubo, Z. Z. Gu, K. Takahashi, Y. Ohko, O. Sato, A. Fujishima, 2002Control of the optical band structure of liquid crystal infiltrated inverse opal by a photoinduced nematic-isotropic phase transition.J. Am. Chem. Soc., 12410950 EOF1 EOF

78. S. Kubo, Z. Z. Gu, K. Takahashi, A. Fujishima, H. Segawa, O. Sato, 2004Tunable photonic band Gap crystals based on a liquid crystal-infiltrated inverse opal structure.J. Am. Chem. Soc., 1268314 EOF9 EOF

79. S. Kubo, Z. Z. Gu, K. Takahashi, A. Fujishima, H. Segawa, O. Sato, 2005Control of the optical properties of liquid crystal-infiltrated inverse opal structures using photo irradiation and/or an electric fieldChem. Mater., 172298 EOF2309 EOF

80. U. Jeong, Y. N. Xia, 2005Photonic crystals with thermally switchable stop bands fabricated from Se@Ag2Se spherical colloids.Angew. Chem., Int. Ed., 443099 EOF103 EOF

81. B. Li, J. Zhou, L. Li, J. Wang, X. H. Liu, J. Zi, 2003Ferroelectric inverse opals with electrically tunable photonic band gap. Appl. Phys. Lett., 83

82. S. Kuai-L, G. Bader, P. V. Ashrit, 2005Tunable electrochromic photonic crystals.Appl. Phys. Lett., 221110

83. J. Khalack, P. V. Ashrit, 2006Tunable pseudogaps in electrochromic WO3 inverted opal photonic crystals. Appl. Phys. Lett., 211112

84. S. Kim, S. Jeon, W. C. Jeong, H. S. Pank, S. Yang, 2008Optofluidic Synthesis of Electroresponsive Photonic Janus Balls with Isotropic Structural ColorsAdv. Mater., 214129 EOF4134 EOF

85. J. Ge, H. Lee, L. He, J. Kim, Z. Lu, H. Kim, J. Goebl, S. Kwon, Y. Yin, 2009Magnetochromatic microspheres: rotating photonic crystals. J. Am. Chem. Soc., 43

86. J. Kim, Y. Song, L. He, H. Kim, H. Lee, W. Park, Y. Yin, S. Kwon, 2011Real-time optofluidic synthesis of magnetochromatic microspheres for reversible structural color patterning.Small, 71163 EOF8 EOF

87. T. D. Schultz, M. A. Rankin, 1985The ultrastructure of the epicuticular interference reflectors of tiger beetles(Cicindela). J. Exp. Biol.,117

88. T. D. Schultz, M. A. Rankin, (1989, 1989Schultz, T.D. & Bernard, G.D. (1989). Pointillistic mixing of interference colours in cryptic tiger beetles. Nature, 337

89. J. Berthier, J. Boulenguez, Z. Balint, 2007Multiscaled polarization effects in Suneve coronata (Lepidoptera) and other insects: application to anti-counterfeiting of banknotesAppl. Phys. A, 1123 EOF130 EOF

90. P. Vukusic, J. R. Sambles, C. R. Lawrence, 2000Colour mixing in wing scales of a butterfly.Nature404457 EOF

91. F. Liu, H. W. Yin, B. Q. Dong, Y. H. Qing, L. Zhao, S. Meyer, X. H. Liu, J. Zi, B. Chen, 2008Inconspicuous structural coloration in the elytra of beetles Chlorophila obscuripennis (Coleoptera).Phys. Rev. E, 012901

92. F. Liu, G. Wang, L. P. Jiang, B. Q. Dong, 2010Structural colouration and optical effects in the wings of Papilio peranthusJ. Opt., 065301

93. J. Vinther, D. E. G. Briggs, J. Clarke, G. Mayr, R. O. Prum, 2010Structural coloration in a fossil featherBiol. Lett., 6

94. A. R. Parker, 2000515 million years of structural colourJ. Opt. A: Pure Appl. Opt., 2R15 EOFR28 EOF

95. M. Kolle, P. M. Salgard-Cunha, M. Scherer, F. M. Huang, P. Vukusic, S. Mahajan, J. J. Baumberg, U. Steiner, (2010, 2010Mimicking the colourful wing scale structure of the Papilio blumei butterfly. Nat. Nanotechnol., 5

96. P. Vukusic, R. Kelly, I. Hooper, 2009A biological sub-micron thickness optical broadband reflector characterized using both light and microwavesJ. R. Soc. Interface, 6S193 EOFS201 EOF

97. J. Huang, X. Wang, Z. L. Wang, 2008Controlled replicaton of butterfly wings for achieving tunable photonic properties. Nano Lett., 10

98. L. P. Biro, K. Kertesz, E. Horvath, G. I. Mark, G. Molnar, Z. Vertesy, J. Tsai, F. Kun, A. Bailint, Z. , J. P. Vigneron, 2009Bioinspired artificial photonic nanoarchitecture using the elytron of the beetle Trigonophorus rothschildi varians as a 'blueprint'J. R. Soc. Interface, 47

99. F. Liu, W. Z. Shi, X. H. Hu, B. Q. Dong, 2012Hybrid structures and the optical effects in Morpho scales with thin and thick coatings using an atomic layer deposition method. Unpublished data.

100. W. Zhang, D. Zhang, T. X. Fan, J. J. Gu, J. Ding, H. Wang, Q. Guo, H. Ogawa, 2009Novel Photoanode Srucure Templated from Butterfly Wing Scales. Chem. Mater., 1

101. R. A. Potyrailo, H. Ghiradella, A. Vertiatchikh, K. Dovidenko, J. R. Cournoyer, E. Olson, 2007Morpho butterfly wing scales demonstrate highly selective vapour responseNat. photon., 1123 EOF128 EOF

102. F. Liu, B. Q. Dong, F. Zhao, X. Hu, X. Liu, J. Zi, 2011Ultranegative angular dispersion of diffraction in quasiordered biophotonic structures Opt. Express, 8

4

THE IMAGING PROPERTIES OF THE METAL SUPERLENS

Chen Zhao[a], Yun-Song Zhou[a,,], Yan Zhang[a], Huai-Yu Wang[b,,]

a Center for Theoretical Physics, Department of Physics, Capital Normal University, Beijing 100048, China

b Department of Physics,Tsinghua University, Beijing 100080, China

ABSTRACT

The zero-refraction property of a metal film can be used to image so that the metal film behaves as a superlens. Experimentally, the metal superlens is sandwiched between the photosensitivity resist (PR) and the dielectric spacer in the photolithograph device to produce images. In this paper, the imaging conditions of the metal superlens, or the parameters of the device containing the superlens are investigated in detail. It is found that the energy zero-refraction effect is the intrinsic property of the metal superlens. The permittivities of the dielectrics at the two sides of the superlens will influence the imaging quality. The rule is drawn for obtaining best images: the two dielectrics should have the same permittivities, so-called symmetry case, and the real part of the permittivity of the metal superlens plus the dielectric permittivity is just equal to 1, i.e.$\epsilon(\mathit{dielectric})+\epsilon_r(\mathit{superlens})=1$.

KEYWORDS

Superlens; Nanolithograph; Imaging; Electromagnetic optics

1. INTRODUCTION

The concept of perfect lens made of double negative (DNG) metamaterials [1], [2],[3] and [4] unlocked a way to obtain an image resolution beyond the traditional diffraction-limit. Because of the difficulties in fabrication of the DNG metamaterials, the applications of the perfect superlens in practice have been baffled. Fortunately, the simplest perfect lens made of silver slab had been proposed in 2000 [2], and demonstrated experimentally in 2005 [5]. After that a lot of works reported the sub-diffractive imaging by metal superlens [6], [7], [8], [9], [10], [11], [12], [13] and [14]. In the demonstration experiment, the silver film was inserted into a photolithographic device and the sub-diffraction pattern was cut on a photosensitive resist (PR). Silver is an epsilon-negative (ENG) material, and this characteristic is essentially different from that of the original perfect lens made of DNG metamaterials. A theoretical analysis showed that [15] when a light beam entered into the ENG or permeability negative (MNG) film with an arbitrary incident angle, the energy flow would penetrate the film with the direction perpendicular to the film surface. This means that the zero-refraction occurs in a single negative film, either ENG or MNG one. As a result, the imaging by a single negative film is caused by the imprint effect, instead of the double focus phenomenon occurring in DNG perfect lens systems. The calculation results based on the zero-refraction principle agreed with the experimental results satisfactorily. Indeed, the zero-refraction has not been touched in standard text books. In the following, when we say a superlens, we mean an ENG or MNG film instead of a DNG perfect lens. In practice, as the superlens works in a complicated photolithograph device, the image quality is determined by several factors, such as the real and imaginary parts of permittivity of the superlens, the properties of the photosensitive resist and medium before the superlens. How these factors influence the image is not very clear yet. In this paper we investigate systematically the influence of the factors. The conclusions are helpful for the design of the photolithograph device in which the superlens works.

2. THE INTRINSIC PROPERTY OF THE SUPERLENS: ZERO-REFRACTION

Although the zero-refraction occurring in superlens has been shown [15], a general and detailed discussion is desired. In this section, we present some new results. Let us consider an incident beam with the width of w illuminating the metal surface, as sketched in Fig. 1(a). The incident electric field is confined in the range from $-x'$ to x', where $2x' = w/\cos\theta$. This pattern will be printed at an observation plane away from the surface by distance z. At this plane, the electric pattern can be expressed as

$$E(x, z) = \frac{1}{2\pi}\int_{-\infty}^{\infty} e^{(ik_x'x+ik_z'z)}dk_x' \int_{-w/(2\cos\theta)}^{w/(2\cos\theta)} te^{i(k_x-k_x')x'}dx' \tag{1}$$

Here t is the transmission coefficient (Fresnel formula), which means the influence of the incident surface on the transmission field. When t =1, the incident beam enters the metal entirely without any reflection. In this case, there is no surface influence and this is the intrinsic characteristics of the superlens. On the contrary, the case of $t\neq 1$involves the surface influence.

First, we discuss the intrinsic characteristics of superlens (t =1). The wavelength of the incident electric field is $\lambda_0 = 365$ nm, at which the permittivity of ϵAg=−2.4012+i0.2488[5]. The width of the incident beam is $w = 3.1609$ μm and the incident angle θ=30°, which results in $x' = w/(2\cos 30°) = 1.825$ μm.

Fig. 1 (b) plots the electric field before and after entering the silver film. The dashed curve expresses the electric field observed on the metal surface, and its envelope line means the amplitude which determines the energy distribution. The solid curve and its envelope line are observed at the plane set at z=35 nm. It is shown that the pattern printed on the incident surface propagates perpendicularly to the observation plane without any lateral shift. In other words, the energy transmission direction is normal to the incident metal surface. The decrease of the envelope line is caused by the damping effect. However, the phase change is along the lateral direction, which can be expressed as $k_1=k_{0x}i$. This result means that the light energy is transmitted with the refraction angle of zero.

The zero-refraction phenomenon in Fig. 1(b) has been noticed in silver superlens [14]. Yet it is unclear whether the zero-refraction is a general phenomenon. Now we investigate this problem in detail and will give a general result. In the following calculation, the structural parameters are taken as that used in Fig. 1(b) but the permittivity can be changed. To do so, we let the real and imaginary parts of the permittivity $\epsilon=\epsilon_r+i\epsilon_i$ vary, so that the metal can be any one, not limited to silver. First, ϵr, the real part of the permittivity is changed but the imaginary part is fixed at that of Ag,ϵ_i=0.2488. The results are displayed in Fig. 1(c). It is seen that when ϵr is negative, the zero-refraction remains.

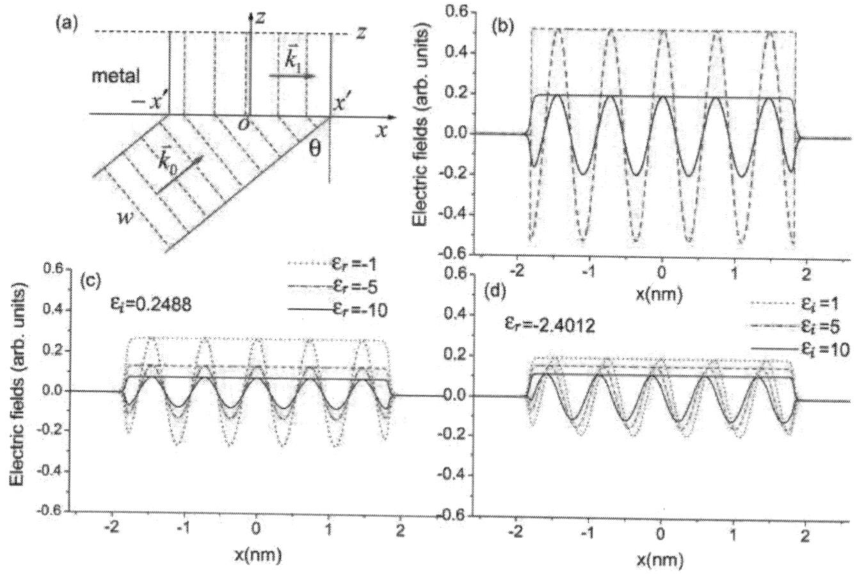

Figure. 1. (a) The sketch of the incident beam illuminating on a metal surface. The dashed lines in the incident beam mean the equiphase surfaces. When the light beam enters the metal, the dashed lines are almost vertical to the metal surfaces. (b) The electric fields on silver front surface (dashed curve) and on the observation plane in the inner of silver (solid curve). The envelope curves show the amplitudes of the corresponding fields. (c) The fields on the observation plane with the structural parameters are the same as those used in (b) but the ϵr is changed. (d) The fields on the observation plane with the structural parameters are the same as those used in (b) but the ϵi is changed.

Now we fix ϵ_r=−2.4012, the real part of the permittivity of Ag and change the imaginary part ϵi. The results are shown in Fig. 1(d). In this case, the equiphase surfaces of the electric field shift in lateral direction. They shift more severely when ϵi is larger. The physical picture is analyzed as follows. The lines presenting the equiphase surfaces are vertical to the film in Fig. 1(a), but they slope to the left in this case, and the wave vector k'has obtained a small z-component. Nevertheless, all the envelope lines do not shift in lateral direction. It is these envelope lines that determine energy transmission directions. So zero-refraction is also valid.

In fact only when ϵ_i=0, the lateral shift of the pattern is exactly zero. In the cases of Fig. 1(b) and (c), the pattern shifts are negligible because the imaginary parts of the permittivities is much smaller compared to the real parts. This conclusion can be understood by the wave vector analysis. Suppose that the permittivity of the metal is$\epsilon_m = \epsilon mr + \epsilon mi$i and the wave number in vacuum is k_0. Inside the metal, the wave number is $k' = k_0\sqrt{\epsilon_m}$ and its tangential component is $k'_x = k_x = k_0 \sin\theta$ when a plane wave illuminates a metal surface with the incident angle θ . So we have $k'^2 = k_0^2\epsilon = k_0^2 \sin^2\theta + k_z'^2$. Here k'_z is the z-component of the wave number in metal, its real and imaginary parts can be expressed as

$$\left.\begin{aligned} k'_{zr} &= k_0\left(\sqrt{(\epsilon_{mr} - \sin^2\theta)^2 + \epsilon_{mi}^2} + \epsilon_{mr} - \sin^2\theta\right)^{1/2} / \sqrt{2} \\ k'_{zi} &= k_0\left(\sqrt{(\epsilon_{mr} - \sin^2\theta)^2 + \epsilon_{mi}^2} - \epsilon_{mr} + \sin^2\theta\right)^{1/2} / \sqrt{2} \end{aligned}\right\} \quad (2)$$

Since ϵmr<0, we have $\epsilon mr - \sin^2\theta < 0$. It is seen from Eq. (2) that as long as ϵmi is a small quantity, $k'_{zr} \to 0$ is irrespective of the incident angle θ. This is the case appearing in Fig. 1(b) and (c). Eq. (2) is deduced from the situation where the incident light is a plane wave. In the case of Fig. 1(a), the incident beam can be expanded by a series of plane wave components with different incident directions. For each of the component, $k'_{zr} \to 0$. Consequently, the fields, as the superposition of these components, shown inFig. 1(b) and (c) do not exhibit the behavior with real wave numbers in z-direction. When ϵi is nonnegligible compared to ϵr, $k'_{zr} \neq 0$. This is the case shown in Fig. 1(d).

It is also seen from Eq. (2) that k'_{zi} is always nonzero. Since it is expressed by the real and imaginary parts of the permittivity, this means that both parts contribute to the damping effect, as shown by the patterns in Fig. 1(c) and (d) where the amplitudes of the patterns are lower than those in Fig. 1(b). From our numerical results, three conclusions about the zero-refraction effect are drawn. The first one is that the energy flow direction perpendicular to the surface of the superlens, which is valid as long as ϵmr<0 andϵmi>0. The second one is that the wave vector is exactly along the surface of superlens when ϵmi=0, and it will get a small perpendicular component when ϵmi increases from zero. The third one is that the energy attenuation comes from both ϵmr and ϵmi.

3. THE DIELECTRIC BEFORE THE SUPERLENS

Let us now discuss the case that the transmission coefficient t≠1, the value of which is to be evaluated by Fresnel formula. Now we use the transfer matrix method to calculate the amplitude distribution (the envelope line) of the electric field on the observation plane. For the dielectric/Ag interface, the results are displayed in Fig. 2 with the parameters the same as those in Fig. 1(b). The four figures (a)–(d) plot the results when ϵr of the dielectric is 1, 2, 2.4, 5, respectively. In the case of vacuum, ϵ_r=1, the envelope line of the electric field is rather different from that in Fig. 1(b). This is because *t* is a function of the lateral wave number *kx*, and the different plane wave components in the incident beam have different transmission coefficients. Subsequently, the superposition of these components on the observation plane cannot recovery the original shape of the incident beam. This property has damaged the zero-refraction effect and will consequently damage the imaging by the superlens. Anyway, with the increase of ϵr, the figures (b)–(d) tend to be closer to the rectangular curve in Fig. 1(b). This is because the large ϵr lead to the result of t→1 (Fresnel formulas). Thus we reach an important conclusion that the dielectric before the superlens should have its permittivity as large as possible to keep the ideal zero-refraction so as to obtain a perfect image.

4. SUPERLENS IN PHOTOLITHOGRAPH DEVICE

Now the superlens is set in the typical photolithography device sketched by the inset inFig. 3. To analyze the imaging, both the incident and exit planes of the superlens should be considered. The incident light with the wavelength λ_0 illuminates the slit cut onto the mask with the width *a*. The superlens is sandwiched between the PR and PMMA layers. The thicknesses of PMMA and the superlens are denoted by d_1 and d_2, respectively. The*x*-axis is along the mask surface and the origin is at the center of the mask slit.

After the incident light passes through the mask slit, it will finally arrive at the PR surface and notch the pattern on it. The notch depth is proportional to $|E(x)_s|^2+|E(x)_p|^2$, the energy strength distributed on the PR surface. Here the $E(x)_s$ and $E(x)_p$ are the electric fields of the *s* and *p* waves at the PR surface respectively. Their expressions are as follows:

$$E_{s,p}(X) = \frac{1}{2\pi}\int_{-\infty}^{\infty} e^{ik'_x x} T_{s,p} dk'_x \int_{-a/2}^{a/2} e^{-ik'_x x'} dx' \tag{3}$$

where the total transmission coefficients for S and P-waves are

$$T_{s,p} = \frac{t_{12}^{s,p} t_{23}^{s,p} e^{i(k_{z1}d_1 + k_{z2}d_2)}}{1 - r_{21}^{s,p} r_{23}^{s,p} e^{2ik_{z2}d_2}} \tag{4}$$

in which

$$t_{i,j}^{s} = \frac{2k_{zi}}{k_{zi} + k_{zj}} \tag{5}$$

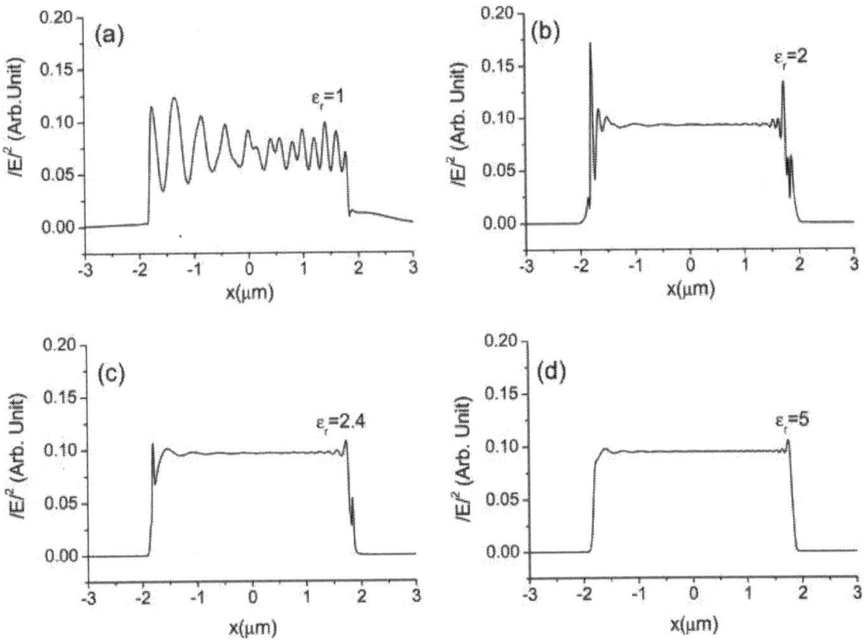

Figure. 2. The amplitude distributions of the electric fields on the observation plane when a silver superlens is covered by the dielectric with the permittivities of ϵ_r=1 (a), ϵ_r=2 (b), ϵ_r=2.4 (c), and ϵ_r=5 (d), respectively.

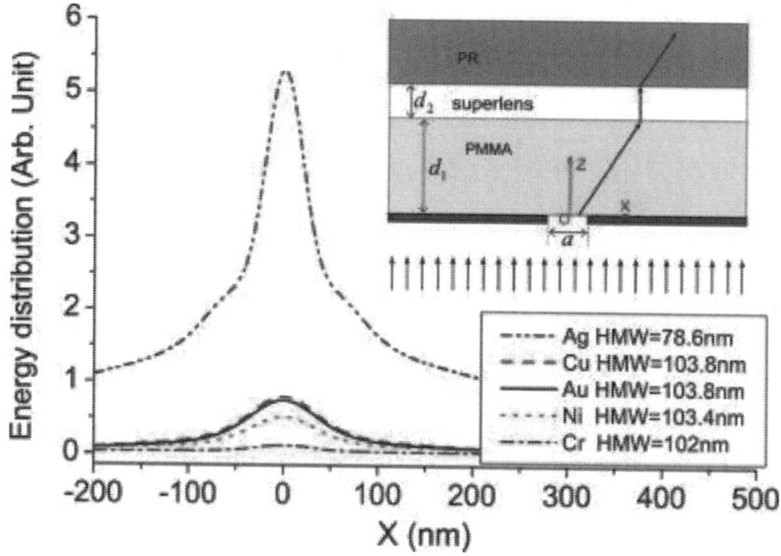

Figure. 3. The energy distributions made by different metal superlens. The inset: a typical photolithograph device.

$$r_{i,j}^{s} = \frac{k_{zi} - k_{zj}}{k_{zi} + k_{zj}} \tag{6}$$

$$t_{i,j}^{p} = \frac{2k_{zi}/(\sqrt{\varepsilon_i}\sqrt{\varepsilon_j})}{k_{zi}/\varepsilon_i + k_{zj}/\varepsilon_j} \tag{7}$$

$$r_{i,j}^{p} = \frac{k_{zi}/\varepsilon_i - k_{zj}/\varepsilon_j}{k_{zi}/\varepsilon_i + k_{zj}/\varepsilon_j} \tag{8}$$

$$k_{zj} = \sqrt{\left(\frac{2\pi}{\lambda_0}\right)^2 \varepsilon_j - k_x'^2} \tag{9}$$

Here i and j=1,2,3, and the characters 1,2,3 denote the mediums PMMA, Ag, and PR, respectively. In other words, t_{12} and r_{21} are the transmission and reflection coefficients at the interface between PMMA and Ag, and t_{23} and r_{23} are that between Ag and PR. k_{zj} can be evaluated by Eq. (9). In our calculation, experimental device parameters [5] are taken, i.e., a=40 nm, $d_1 = 40$ nm, $d_2 = 35$ nm. The incident wavelength is $\lambda_0 = 365$ nm, and the corresponding permittivities of PMMA and PR areϵ_1=2.3013+i0.0014 and ϵ_{PR}=1.5170+i0.00046, respectively. Hereafter, the two permittivities are called device parameters.

The energy distributions on the PR surface under the device parameters are displayed inFig. 3. The sandwiched superlens is one of Ag, Cu, Au, Ni and Cr. Their permittivities corresponding to the incident wavelength $\lambda_0 = 365$ nm are ϵ_{Ag}=−2.4012+i0.2488[5],ϵ_{Cu}=−2.1896+i4.953[16], ϵ_{Au}=−0.52239+i6.39038[17], ϵ_{Ni}=−2.0845+i7.0308[17], ϵ_{Cr}=−8.55,+ i8.96[5].

The right bottom corner of Fig. 3 lists the half-maximum width (HMW) of the energy distributions curves. It is seen that among all the superlens, Ag superlens gives the highest peak with the narrowest HMW, 78.8 nm. The peak of the curve in Fig. 3, or the depth of the notch, can be controlled by incident energy. To obtain a good image, the notch on the PR should have narrow HMW and clear edge as far as possible. Therefore, we concentrate to investigate these two points. In doing so, the real part ϵr and imaginary part ϵi of the permittivity of the superlens are varied in large ranges, and various energy distribution curves are gained. It is found that besides the Gauss-like distributions, there are also more complicated curves when the permittivity is in some regions, which will be shown in Fig. 4 and Fig. 5 and discussed later. The complicated energy distributions have wider HMWs and will cause ambiguous edges of the notch. So the Gauss-like energy curves are preferred. Gauss curve is referred to as the normal curve or normal shape. For example, the Cu-curve in Fig. 3 is closer to Gaussian curve than the Ag-curve. Although the latter has a narrower HMW, the former has clearer edges.

It has been concluded in Section 3 that the large permittivity of the dielectric before the superlens is in favor of the imaging effect. This property can be used to adjust the image quality. The superlens is made by silver and the device parameters are as above except that the real part of the permittivity of the PMMA ($\epsilon_r(PMMA)$) is changed. Fig. 4 plots the calculation results when $\epsilon_r(PMMA)$=1, 2, 3, 4, 7, 10 and 15. The curve of$\epsilon_r(PMMA)$=1 shows the complicated shape and that of $\epsilon_r(PMMA)$=2 is improved but still has little bell mouth on the bottom. The curve of $\epsilon_r(PMMA)$=3 exhibits the normal shape and the HMW is 55.6 nm, narrower than that in Fig. 3 (HMW=78.6 nm). As$\epsilon_r(PMMA)$ is further increased, all the curves show the normal shape, but their peaks lower gradually, and as a result, the HMW becomes wider. This demonstrates that a too large $\epsilon_r(PMMA)$ will cause a wider line width, although it may yield clear edges. Thus, to take into account both the clear edges and narrowest line width of the imaging, an optimum $\epsilon_r(PMMA)$ should be chosen.

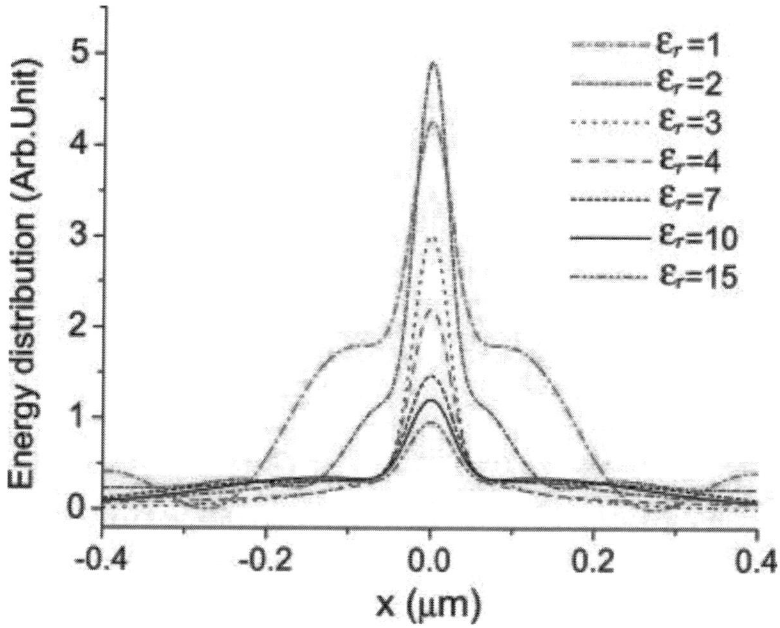

Figure. 4. Energy distribution curves on the PR surface in a device containing a silver superlens under device parameters except that the real part of the permittivity of PMMA (ϵr) is changed.

Naturally, one would ask how to find the optimum $\epsilon_r(PMMA)$. In Fig. 4, the transition from the complicated curve to the normal curve occurs between $\epsilon_r(PMMA)=2$ and 3. It is noticed that the real part of the superlens permittivity is $\epsilon_r(Ag)=-2.4012$, the absolute value of which is just between 2 and 3. On the other hand, our calculation results show that when $\epsilon_r(PMMA)=\epsilon_r(PR)$ (symmetry case), the normal energy curve is easily produced. Therefore, we keep the device in the symmetry cases and with device parameters, but the real parts of permittivities of the metal superlens and that of PMMA and PR are changed. The calculated energy distribution curves are shown in Fig. 5.

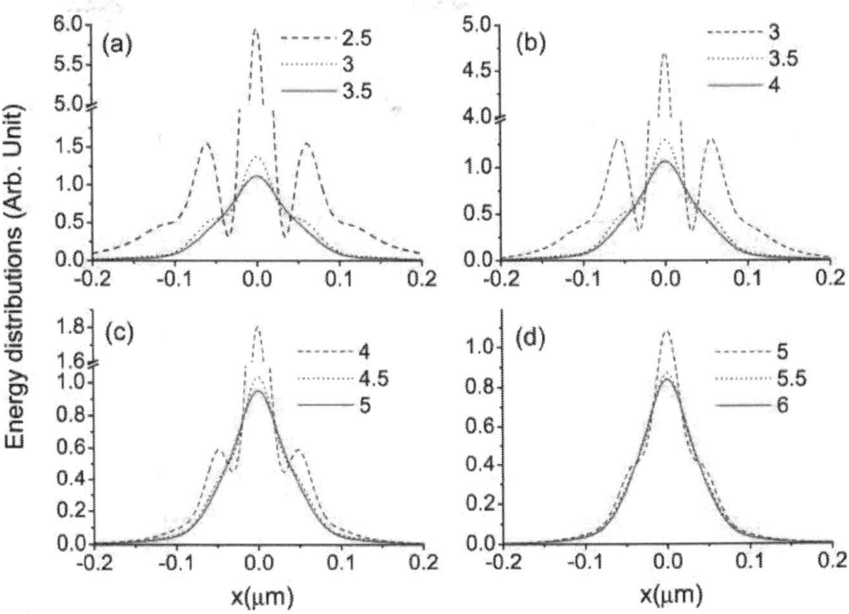

Figure. 5. Energy distribution curves on PR surface under device parameters but the real parts of permittivities of the PMMA, PR and superlens are modified. The device keeps the symmetry case $\epsilon_r(PMMA)=\epsilon_r(PR)$, and the values of which mark the lines. The real part of the superlens permittivity is −2.4012(Ag) in (a), −3 in (b), −4 in (c), and −5 in (d).

In Fig. 5(a) the superlens is a silver film. When the permittivity of the dielectric is $\epsilon_r=3.5$, the energy curve shows the normal distribution. In this case, $|\epsilon_r(superlens)|=|\epsilon_r(Ag)|=2.4012$. That is to say, the transition between the complicated and normal curves occurs at about $\epsilon_r(T)=|\epsilon_r(superlens)|+1$ which is considered as a rule. InFig. 5(b), $|\epsilon_r(superlens)|=3$ and the transition occurs at $\epsilon_r=4$. Thus it follows the rule that $\epsilon_r(T)=|\epsilon_r(superlens)|+1$. In Fig. 5(c) and (d), $|\epsilon_r(superlens)|=4$ and 5, respectively, and the transition curves occur when $\epsilon_r=5$ and 6, respectively. Therefore, the rule $\epsilon_r=|\epsilon_r(superlens)|+1$ remains valid. In conclusion, the permittivity matching rule is $\epsilon_r(T)+\epsilon_r(superlens)=1$. Since the PR and PMMA are dielectrics, the imaginary parts of the permittivity are negligible. As a result, the rule can be expressed as $\epsilon(dielectric)+\epsilon_r(superlens)=1$.

5. SUMMARY

We investigate in detail the optimal conditions of the imaging devices comprising a superlens by zero-refraction. Our conclusions are outlined as follows. The zero-refraction effect is an intrinsic property of metal. The dielectric covering the incident surface of the superlens will influence the imaging quality, and its value should be as large as possible to produce good images in the inner of the superlens. In the photolithography devices, to obtain the best images requires the so-called symmetry case and the matching rule: the permittivities of dielectrics covering both surfaces of the superlens should be the same and the permittivities observe the rule $\epsilon(\textit{dielectric})+\epsilon_r(\textit{superlens})=1$.

ACKNOWLEDGMENT

This work is supported by the 973 Program of China (Grant no. 2013CBA01702) and the National Natural Science Foundation of China (Grant nos. 11274233 and 61275028).

REFERENCES

1. V.G. Veselago **The electrodynamics of substances with simultaneously negative values ϵ and μ** Sov. Phys. Usp., 10 (1968), pp. 509–514

2. J.B. Pendry **Negative refraction makes a perfect lens** Phys. Rev. Lett., 85 (18) (2000), pp. 3966–3969

3. D.R. Smith, J.B. Pendry, M.C.K. Wiltshire **Metamaterials and negative refractive index** Science, 305 (2004), pp. 788–792

4. Tran Minh Hien, Ho Trung Dung **Image of an emitting dipole by a superlens** Phys. Rev. A, 85 (2012), p. 015804

5. N. Fang, H. Lee., C. Sun, X. Zhang **Sub-diffraction-limited optical imaging with a silver superlens** Science, 308 (2005), p. 534

6. Z. Shi, V. Kochergin, F. Wang **Depth-of-focus (DoF) analysis of a 193 nm superlens imaging structure** Opt. Express, 17 (22) (2009), pp. 20538–20545

7. J.J. Dong, J. Liu, X.X. Zhao, P. Liu, J.H. Xie, Y.T. Wang **Image distance shift effect of the metal superlens and its applications to photolithography** Europhys. Lett., 102 (2013), p. 24002

8. C. Jeppesen, R.B. Nielsen, A. Boltasseva, S. Xiao, N.A. Mortensen, A. Kristensen **Thin film Ag superlens towards lab-on-a-chip integration** Opt. Express, 17 (25) (2009), pp. 22543–22552

9. P. Chaturvedi, W. Wu, V.J. Logeeswaran, Z.N. Yu, M.S. Islam, S.Y. Wang, R.S. Williams, N.X. Fang **A smooth optical superlens** Appl. Phys. Lett., 96 (2010), p. 043102

10. Fuyang Xu, Genhua Chen, Chinhua Wang, Bing Cao, Yimin Lou **Superlens imaging with a surface plasmon polariton cavity in imaging space** Opt. Lett., 38 (19) (2013), pp. 3819–3822

11. Y. Zhang, M.A. Fiddy **Covered image of superlens** Prog. Electromagn. Res., 136 (2013), pp. 225–238

12. J. Liu, W. Xia, S.Q. Zhang, Q.M. Dong, X.W. Guo, Z.Y. Zhang **Subwavelength imaging of self-assembled triangular array through a silver superlens** Microelectron. Eng., 123 (2014), pp. 28–32

13. S. Dai, Q. Ma, T. Andersen, A.S. Mcleod, Z. Fei, M.K. Liu, M. Wagner, K. Watanabe, T. Taniguchi, M. Thiemens, F. Keilmann, P. Jarillo-Herrero, M.M. Fogler, D.N. Basov **Subdiffractional focusing and guiding of polaritonic rays in a natural hyperbolic material** Nat. Commun., 6 (2015), p. 6963

14. L. Fang, N. Yao, M.B. Pu, C.T. Wang, X.G. Luo **Sub-diffraction-limited magnified Talbot imaging in cylindrical metamaterial** Appl. Phys. A, 118 (2015), pp. 1543–1549

15. Y.S. Zhou, H.Y. Wang, H. Wang **Zero refraction in natural materials and the mechanism of metal superlens** Europhys. Lett., 98 (2012), p. 67005

16. H.J. Hagemann, W. Gudat, C. Kunz **Optical constants from the far infrared to the x-ray region: Mg, Al, Cu, Ag, Au, Bi, C, and Al_2O_3** J. Opt. Soc. Am., 65 (17) (1975), pp. 742–744

17. D. Aleksandar Rakić, B.D. Aleksandra, M.E. Jovan, L.M. Marian **Optical properties of metallic films for vertical-cavity optoelectronic devices** Appl. Opt., 37 (1998), pp. 5271–5283

5

FLEXIBLE OPTICAL WAVEGUIDE BENT LOSS ATTENUATION EFFECTS ANALYSIS AND MODELING APPLICATION TO AN INTRINSIC OPTICAL FIBER TEMPERATURE SENSOR

Mustapha Remouche, Francis Georges, Patrick Meyrueis

Laboratoire des Systèmes Photoniques, École Nationale Supérieure de Physique de Strasbourg, Université de Strasbourg, Strasbourg, France

ABSTRACT

The temperature dependence of the bending loss light energy in multimode optical fibers is reported and analyzed. The work described in this paper aims to extend an initial previous analysis concerning planar optical waveguides, light energy loss, to circular optical waveguides. The paper also presents a novel intrinsic fiber optic sensing device base on this study allowing to measure temperatures parameters. The simulation results are validated theoretically in the case of silica/silicone optical fiber. A comparison is done between results obtained with an optical fiber and the results obtained from the previous curved optical planar waveguide study. It is showed that the bending losses and the temperature measurement range depend on the curvature radius of an optical fiber or waveguide and the kind of the optical waveguide on which the sensing process is implemented.

KEYWORDS

optical fiber; losses; curvature; sensor; temperature; micro technology

1. INTRODUCTION

A curvature effect is easily reached with optical fibers; therefore many laboratories have investigated the effects of curvatures on optical fiber measuring responses. Gratings implemented in optical fibers provide measuring performances related to bending that were also investigated [1-12]. Therefore, the power attenuation coefficient of bent fibers is one of the parameters that must be determined for using the fiber as a transducer.

In previous paper [1], we described a geometrical method to determine the local numerical aperture and the light output power attenuation with the bending of an optical waveguide. We showed that this method can be applied only when the optical waveguide is curved during the manufacturing at a temperature close to glass melting. In this case, the core and the cladding refractive index are modified by temperature effects independently of the curvature radius.

In this a previous work [1], we described a set of methods for using a flat optical waveguide as a transducer for measuring temperature by bending, in which only some transmitted light intensity effects are involved. We also investigated previously some photonic effects allowing using a multimode integrated optical waveguide as an intrinsic temperature sensor operating by light intensity modulation at the output of the sensor.

The purpose of this new work is to extend the analysis of the planar optical waveguide response to the temperature [1], to a circular optical waveguide.

An optical fiber is a good example of cylindrical optical waveguide. The optical fiber bending loss phenomena is used as a transduction effect in some types of intrinsic optical fiber sensors (temperature, displacement, strain...) [1,2,13-20].

The step-index optical fiber that we use is curved during its manufacturing at a high temperature. A geometrical modeling is used to describe the light propagation in the optical fiber and to determine the light power attenuation at the output of the fiber due to the fiber bending. In this case the geometrical approach is similar to the one presented for the planar waveguide sensing modeling.

2. ANALYSIS IF THE POWER ATTENUATION WITH THE BENDING OF AN OPTICAL FIBER

The guidance of the core rays in a straigth step-index optical fiber is achieved by ensuring that the propagation angle q, satisfies the condition: $0 \le \theta \le \theta_c$, where the critical angle, q_c, and the critical angle, a_c, are given respectively, at room-temperature (T_0 = 20°C) by:

$\theta_c = \cos^{-1}(n_2/n_1)$ and $\alpha_c = \sin^{-1}(n_2/n_1)$. The numerical aperture (NA) is given at T_0 by:

$NA = n_1 \sin\theta \le \left(n_1^2 - n_2^2\right)^{1/2}$. n_1 and n_2 are the core and the cladding refractive index respectively [1].

When the optical fiber is bent with a curvature radius R (figure 1), the local numerical aperture at a given location of the optical fiber curved part will be changed. In a meridional-plane of the optical fiber, when the position angle at the beginning of the bend is ϕ = 0° or 180°, the optical fiber behavior becomes identical to the one an optical planar waveguide [1], and the local numerical aperture NA_1 is given, at T_0, by:

$$NA_l\left(R,\rho_0\right) = n_1 \sin\theta = n_1^2\left[1 - \frac{n_2^2}{n_1^2}\left[\frac{R+\rho}{R+\rho_0}\right]^2\right]^{1/2} \tag{1}$$

where r is the fiber core radius and $r(=R+\rho_0)$ is the abscissa on the input optical fiber aperture where the origin is "O". "r_0" will satisfy the relation: $-\rho \le \rho_0 \le \rho$. However, in all others optical fiber planes, when $\phi \ne 0^\circ$ and 180°, the local numerical aperture, NA_l, can be calculated by using Equation (1), where the quantity $(R-\rho_0 \cdot \cos\phi)$ replaces $(R+\rho_0)$

and "ρ_0" to satisfy the relation: $0 \le \rho_0 \le \rho$. Finally, the local numerical aperture of a curved step-index optical fiber is given by:

$$NA_l\left(R,\rho_0,\phi\right) = n_1^2\left[1-\frac{n_2^2}{n_1^2}\left[\frac{R+\rho}{R-\rho_0\cdot\cos\phi}\right]^2\right]^{1/2} \quad (2)$$

where ϕ is required to satisfy the relation: $0 \le \phi \le 2\pi$.

Equation (2) becomes identical to equation: $NA = \left(n_1^2 - n_2^2\right)^{1/2}$ when R is infinite (straight optical fiber case) and becomes identical to equation (1) when f = 0° or 180° (bent step-index optical planar waveguide case). We used the silica/silicone step-index optical fiber with the following characteristics at T_0, for a wavelength l = 633 nm: n_1 = 1.4570, n_2 = 1401, 2r = 200 µm and NA = 0.4000.

In **Figure 2**, we plot the local numerical aperture, at T_0 for various values of R, as a function of the position "r_0" with a value of f. It is clearly seen that the local numerical aperture increase with the increase of the values of R and r_0. The maximum of the local numerical aperture is obtained when the angle f is equal to 180°.

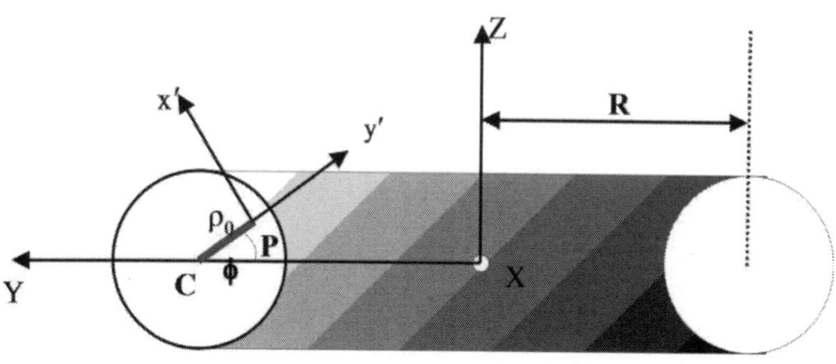

Figure 1. Schematic presentation of a section of a fiber bent with an arc of radius R.

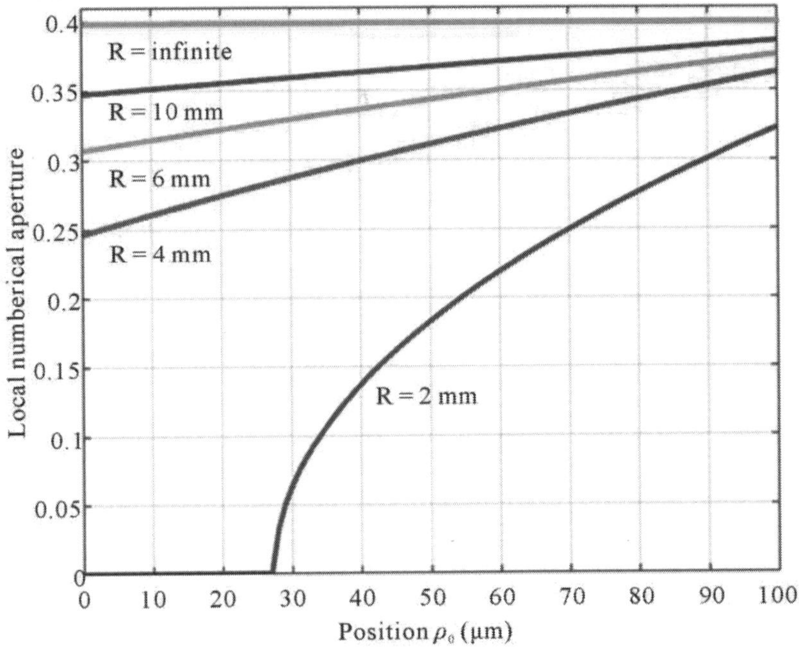

Figure 2. Local numerical aperture according to the position "r_0" for several values of R with f = 135°.

3. AN APPROACH OF THE DESCRIPTION OF THE LIGHT PROPAGATION IN A BENT OPTICAL FIBER

The ray paths in the core of a step index fiber are straight lines, but the geometrical description is more complicated than in a planar waveguide, that we analyzed previously [1], due to the presence of the skew rays and of the depending on when the optical fiber is curved. In the bent planar waveguide [1], the angle of incidence of a ray on the outer core-cladding interface remains the same along a given ray path, and this property is true for all the ray paths. However, in the bent optical fiber, only the rays entering the bent part of the fiber in the meridionalplane containing this bent part behave in this manner. For the skew rays entering this plane, the subsequent reflections within the core do not follow a simple repeatable pattern because of the asymmetry introduced by bending the fiber. The complicated form of the ray invariants [3] and the differential equations describing the ray path offer no possibility of a simplification. We therefore propose a numerical technique for tracing each ray separately along the bent fiber.

In the **Figure 3**, we consider an arbitrary incident ray entering the bent section of the optical fiber from the cross section X'X at point "P" at the beginning of the bent part (x = 0). The bent part of optical fiber forms a torus portion by the revolution of the circular cross section about the origin "O" with a radius R. The axis system OXYZ has an axis OX pointing directly into the paper. A local coordinate system ox'y'z' with oz' pointing into the paper is also indicated at P. An arbitrary ray, incident at "P", has its direction cosines ($\cos\theta_{x'}, \cos\cos\alpha_i, \cos\cos\theta_{z'}$) relative to ox'y'z'. The vector OP is given by:

$$\boldsymbol{OP} = \left(0, R - \rho_0 \cos\phi, \rho_0 \sin\phi\right) \tag{3}$$

where the distance of the point "P" from the fiber center is r_0, if we let r be a point along the ray distant "L" from "P" then:

$$\begin{aligned} \boldsymbol{r} &= L\cos\cos\theta_{z'}\hat{X} \\ &+\left[R - \rho_0\cos\phi + L\left(\cos\cos\theta_{x'}\sin\phi - \cos\alpha_i\cos\phi\right)\right]\hat{Y} \\ &+\left[\rho_0\sin\phi + L\left(\cos\phi + \cos\alpha_i\sin\phi\right)\right]\hat{Z} \end{aligned} \tag{4}$$

If the ray meets the torus at "Q_1" then we can write $|\boldsymbol{L}| = \overline{PQ_1}$, where "$Q_1$" is the position of the first reflection. At "Q_1", we have:

$$\left|\boldsymbol{r} - R\sin\xi\hat{X} - R\cos\xi\hat{Y}\right| = \rho \tag{5}$$

And this gives the angular distance x around the axis of the bent optical fiber curvature:

$$\tan\xi = \boldsymbol{r}\hat{X}/\boldsymbol{r}\hat{Y} \tag{6}$$

x_1 corresponds to $\overline{POQ_1}$ angle. From Equation (6), we deduce an equation which will provide, as a function of the position "L", the intersection of the optical ray with the bent optical fiber core-cladding interface. The smallest real positive solution of the equation represents the distance "L_1".

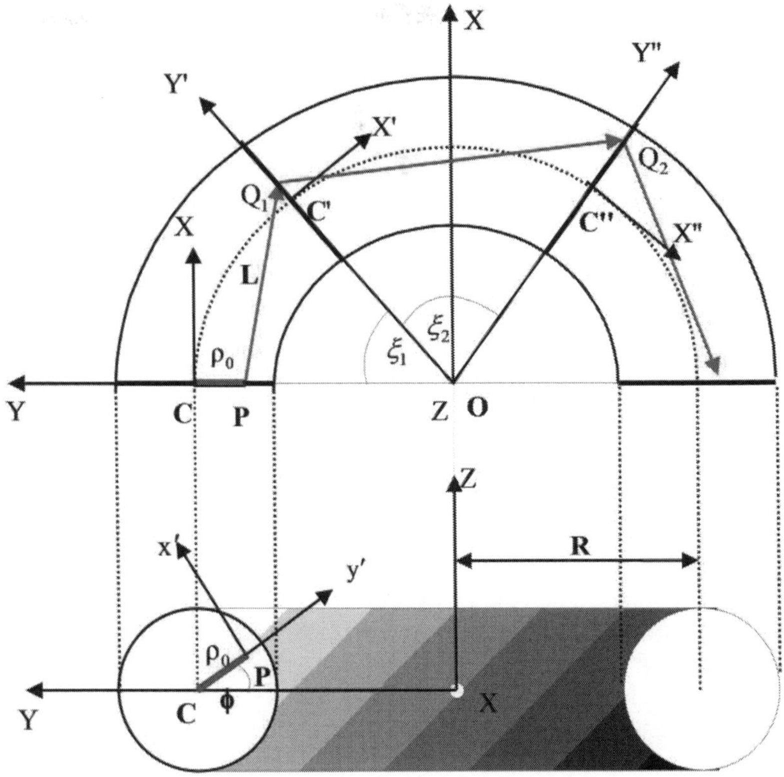

Figure 3. Schematic presentation of a section of a fiber bent with the curvature following an arc having a radius R.

From the point of the first reflection "Q_1" on the optical fiber torus, the optical ray is reflected to another point on the surface of the torus (core-cladding interface) around the bend.

After the first reflection all solutions for the distance "L" to the next reflection are given as solutions of the cubic problem. For obtaining the cubic equation, we must replace r_0 by r in the quadratic equation.

Some simple coordinates rotations and translations simplify the calculation of the incident and the reflected ray angles at each reflection point [2]. The reflected ray direction cosines at "Q_1" are given by multiplying a matrix transformation "M" by the direction cosines a "P" where:

$$\begin{bmatrix} \text{Direction} \\ \text{cosines} \\ \text{at } Q_1 \end{bmatrix} = [M] \cdot \begin{bmatrix} \cos\theta_{x'} \\ \cos\alpha_i \\ \cos\theta_{z'} \end{bmatrix} \tag{7}$$

With: (see Equation (8)).

$$M = \begin{bmatrix} \cos\zeta\cos\phi + \sin\zeta\cos\xi\sin\phi & \sin\phi\cos\zeta - \sin\zeta\cos\xi\cos\phi & \sin\xi\sin\zeta \\ \sin\zeta\cos\phi - \cos\zeta\cos\xi\sin\phi & \sin\zeta\cos\phi + \cos\zeta\cos\xi\cos\phi & -\sin\xi\cos\zeta \\ -\sin\phi\sin\xi & \sin\xi\,\text{coss}\,\phi & \cos\xi \end{bmatrix} \tag{8}$$

A rotation around "Q_1" will bring the local x-axis tangential to the surface of the torus at Q_1. This rotation is z. All the others parameters were defined previously.

Having determined the geometry of the ray path, we can then calculate the fractional power loss at each reflection point along a given path by using the following equation:

$$P(\xi) = P(0)\exp(-\gamma\xi) \tag{9}$$

where g is the attenuation coefficient of each ray, which varies from one reflection to the next one along the bent optical fiber, it is given by:

$$\gamma = \sum_{i=1}^{N} T_i \Big/ \sum_{i=1}^{N} \Delta\xi_i \tag{10}$$

where $\Delta\xi$ is the angular separation between two successive reflections and N is the total number of reflections. We can then use the Generalized Fresnel's Law to calculate the transmission coefficient T_i at each reflection point along a given path.

An algebraic expression for the transmission coefficient of the refracted rays ($V \ll 1$), is given by [3,11,12]:

$$T_i = \frac{4K_{y_1'}K_{y_2'}}{\left(K_{y_1'} + K_{y_2'}\right)^2} \tag{11}$$

where $K_{y_1'}$ and $K_{y_2'}$ are given by: $K_{y_1'} = k_1 \cos\alpha_i$ and $K_{y_2'} = k_1\left(\sin^2\alpha_c - \sin^2\alpha_i\right)^{1/2}$ and: $k_1 = 2\pi n_1/\lambda$.

For tunneling rays when $V \gg 1$, the transmission coefficient at a reflection point of the radius ρ_c is given by:

$$T_i = \frac{4K_{y_1'}K_{y_2'}}{\left(K_{y_1'} + K_{y_2'}\right)^2}\exp\left\{-\frac{4}{3}V^{3/2}\right\} \tag{12}$$

where the parameter "V" is given by:

$$V = \frac{\rho_c}{2}\left(1 - \frac{\sin^2\alpha_c}{\sin^2\alpha_i}\right)\left(\frac{2k_1^2\sin^2\alpha_i}{\rho_c}\right) \tag{13}$$

The parameter ρ_c is the radius of curvature of the core-cladding interface in the incidence plane at "0". It is defined by the normal to the interface and the incident ray direction. It is given by [3,11,12]:

$$\rho_c = \frac{\rho_{x'}\rho_{z'}\sin^2\alpha_i}{\rho_{x'}\cos^2\theta_{z'} + \rho_{z'}\cos^2\theta_{x'}} \tag{14}$$

where $\rho_{x'} = \rho$ and $\rho_{z'} = \rho + R/\cos\phi$.

When $V \cong 0$, the transmission coefficient is givenby:

$$T_i = 3.182\left(\frac{\lambda\cos^2\theta_c}{2\pi n_1\rho_c}\right)\frac{1}{\sin\theta_c} \tag{15}$$

But there are some optical rays that reach the interface with incidence angles close to θ_c.

Finally, the total intensity at the end of the angular length "ξ" of the bent part of the optical waveguide is found by a quadruple summation of Equation (9) across the cross-sectional area (r,f) of the optical fiber at X'X and the distribution of the ray angle (q,y) [3,11,12]:

$$P_s(\xi) = P_0 \int_0^{\varrho} r\mathrm{d}r \int_0^{2\pi} \mathrm{d}\psi \int_0^{\theta_c} \cos\theta \sin\theta \mathrm{d}\theta \int_0^{2\pi} \mathrm{d}\phi \exp(-\gamma\xi) \tag{16}$$

We plot in **Figure 4** the normalized power attenuation against the normalized distance (z/r) along the bent fiber axis for a step index optical fiber. Where "z" is the curvature length (z = R·x). Each curve shows the same characteristics. Initially, there is the transition region, having a rapid power loss. This region dominated by the refracted rays and tunnel rays with the largest attenuation coefficient. The behavior of the bent part of the fiber is similar to the planar waveguide bent part [1], but the loss of power in the transition region is not as important, since only a few rays close the meridional plane have large losses.

In **Figure 5**, we plot the normalized power attenuation against normalized distance (z/r) along the bent planar waveguide and along the bent optical fiber for the same curvature radius (R = 2 mm). Beyond the transient region, the loss of power for a bent optical fiber case is higher than for the bent planar waveguide. This is because the skews rays can pass through the regions of high attenuation, although they may initially have a low attenuation.

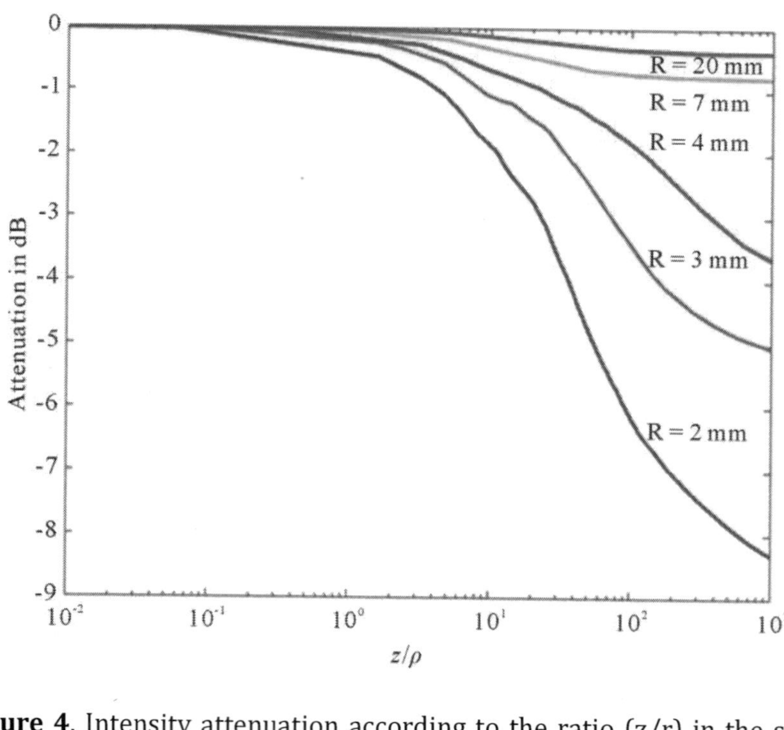

Figure 4. Intensity attenuation according to the ratio (z/r) in the step profile optical fiber for several values of R.

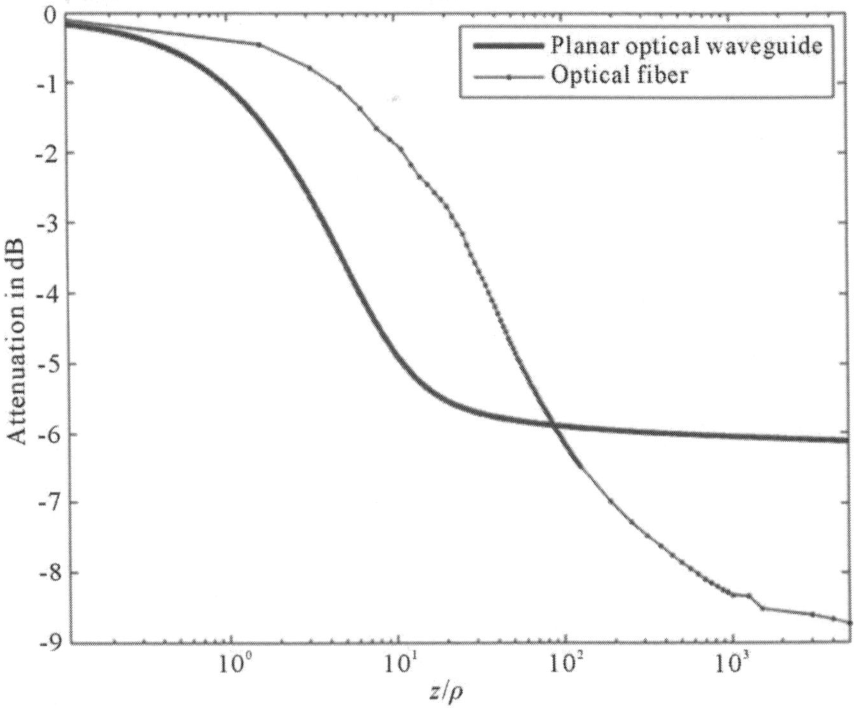

Figure 5. Intensity attenuation comparison between a bent planar waveguide and the bent optical fiber for the same curvature radius (R = 2 mm).

4. TEMPERATURES MEASUREMENTS PRINCIPLE BY USING A BENT OPTICAL FIBER AS A TRANSDUCER

The optical fiber sensor that we propose is an intrinsic optical fiber sensor; the sensitive element is the curved part of the fiber (**Figure 6**). This sensor principle is based on the variations of the sensing fiber output light intensity according to the fiber temperature variations. The light energy losses induced by the bending effect of the fiber will be compensated dynamically by the thermal effect; consequently, the output signal intensity is modulated only by the temperature variations.

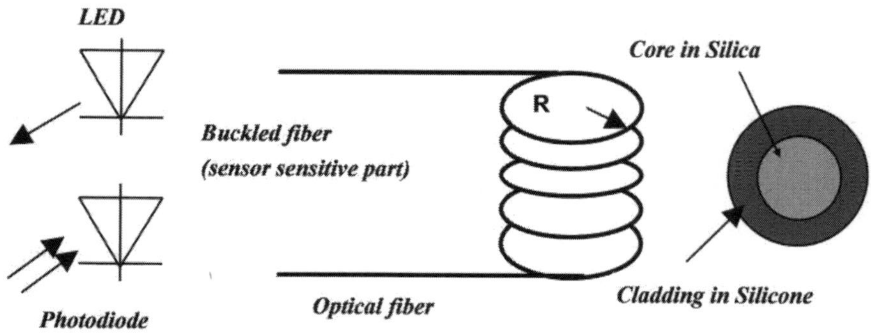

Figure 6. Temperature sensor general diagram.

5. THE TEMPERATURE EFFECT ON A LOCAL NUMERICAL APERTURE INSIDE A BENT OPTICAL FIBER

For a given radius (R = R_0) of the optical fiber curvature and from equation (2), the local numerical aperture inside this bent portion of the fiber according to the temperature is written in the following way:

$$NA_l\left(R_0,T,\rho_0,\phi\right)$$
$$=n_1^2\left(R_0,T\right)\left[1-\frac{n_2^2\left(R_0,T\right)}{n_1^2\left(R_0,T\right)}\left[\frac{R_0+\rho}{R_0-\rho_0\cos\phi}\right]^2\right]^{1/2} \tag{17}$$

where the core refractive index and the cladding refractive index are written according to the temperature (T) in the following way [1]·

$$n_1\left(R_0,T\right)=n_1\left(T_0\right)+K_1\left(T-T_0\right),$$
$$n_2\left(R_0,T\right)=n_2\left(T_0\right)+K_2\left(T-T_0\right) \tag{18}$$

The coefficient $K_1=\mathrm{d}n_1/\mathrm{d}T$ and $K_2=\mathrm{d}n_2/\mathrm{d}T$ are respectively the thermo-optic coefficient of core the refractive index and the thermo-optic coefficient of cladding the refractive index. $K_1 = -3.78 \times 10^{-4}/°C$ and $K_2 = 1.7744 \times 10^{-5}/°C$.

Silicone-based polymers possess a unique set of properties that makes them highly suitable for optical applications. The excellent thermal stability (−115°C to 260°C) allows this material to be useful for high temperature sensing applications [22-26].

In this optical fiber the small positive thermo-optic coefficient effect in inorganic glass waveguides used as the core is canceled out by using the negative thermooptic coefficient of polymers used to constitute the cladding.

For an applied temperature T, we plot in figure 7 the local numerical apertures in an optical fiber according to the value of "ρ_0". We observe that the local numerical aperture increases when the applied temperature increases. Consequently, an optical ray unguided at room temperature becomes guided at temperature greater than T_0.

6. BENT OPTICAL FIBER TEMPERATURE RESPONSE AS A SENSOR

In this section, we analyze the effect of the temperature variations on the light propagation in a curved optical fiber. We present the effects of the curvature radius on the optical fiber temperature response. For this analysis, the curvature radius R and the length of the bent part of the optical fiber (x) are given. The refractive index of the core and the cladding of the optical fiber depend on the temperature. The geometrical model is used to evaluate the light output power according to the temperature. The output power at the end of the transducer bent part of the fiber is given by:

$$P_s(R_0,T) = P_0 \int_0^{\rho} r\mathrm{d}r \int_0^{2\pi} \mathrm{d}\psi \int_0^{\theta_c} \cos\theta \sin\theta \mathrm{d}\theta \int_0^{2\pi} \mathrm{d}\phi \exp\left(-\gamma(R_0,T)\xi\right) \quad (19)$$

For a given curvature length (x = 2pR_0), we plot in **Figure 8** the normalized output light intensity response of the sensor to the temperature variations for several values of R_0.

The response curve of the optical fiber operating as a temperature sensor that we propose in **Figure 8** is similar to the one of a planar waveguide temperature sensor [1]. For this first approach we have not considered errors measuring protocols. We note that the transmission rate (P/P_0) increases when the local numerical aperture is increasing up to a saturation level. Each curve is corresponding to the recovery of the losses induced by the refracted optical rays and the recovery of the losses induced by the tunnel effect, up to saturation. We can say that the temperature sensing range of the temperature optical fiber that we propose depends on the curvature radius of the optical fiber.

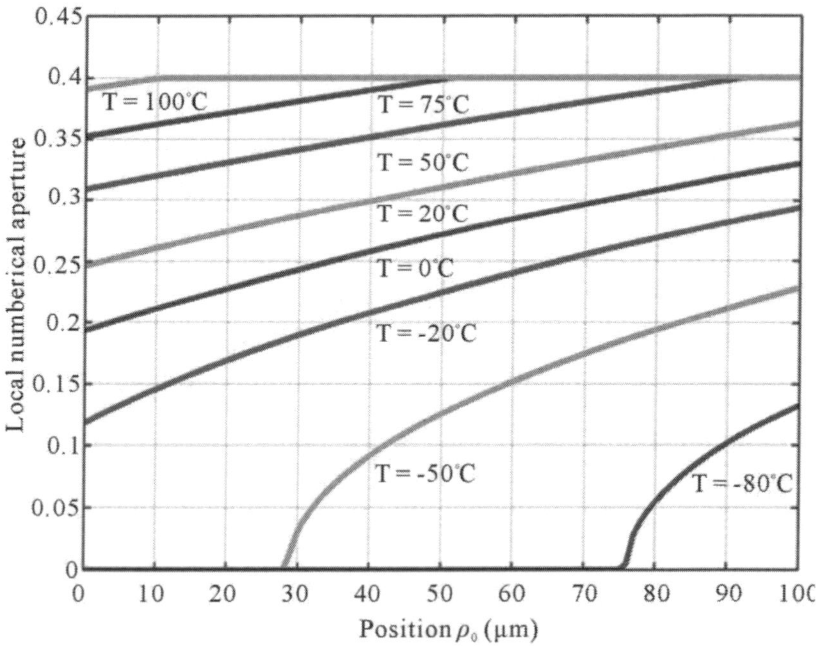

Figure 7. Local numerical aperture according to the position "r_0" for several applied temperature, with R_0 = 4 mm and f = 135°.

We deduce from **Figure 8** the following statements:

- The part of the response curve between Tc1 and T_0 is corresponding to intensity losses caused only by the temperature effect.
- The part of the response curve between T_0 and Tc2 is corresponding to intensity losses caused only by the optical fiber curvature.

- The linear region of the optical fiber temperature sensor response is increasing when the curvature radius of the bent fiber is decreasing.

The sensor sensitivity is given by:

$$S = \frac{dP(R_0, T)}{dT} \tag{20}$$

where $P(R_0,T)$ is the output intensity. We deduce from **Figure 8** that the sensitivity of the temperature sensor is inversely proportional to the curvature radius.

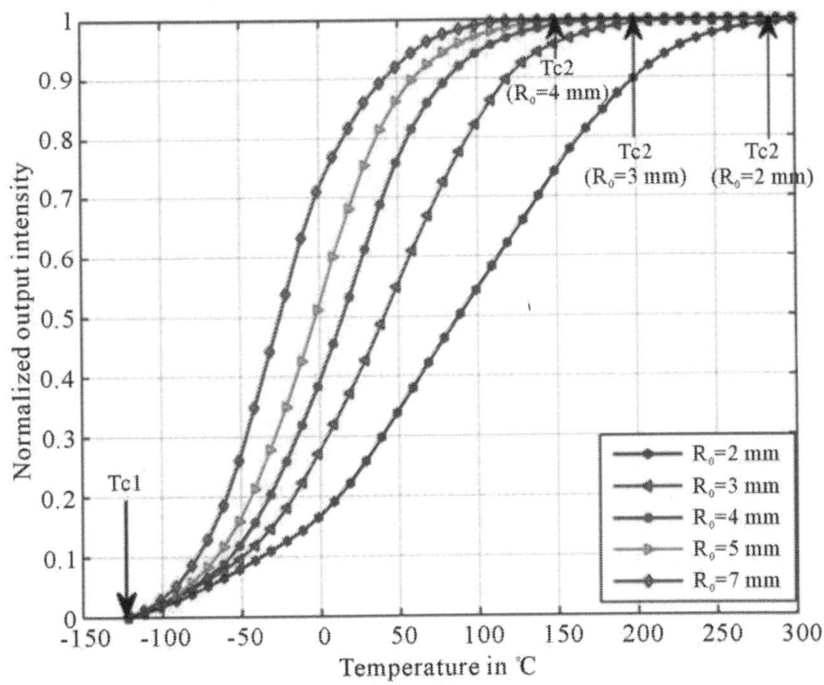

Figure 8. Transmission rate according to the temperature for various curvature radiuses.

Figure 9 shows the temperature response curve fitting for $R_0 = 2$ mm. In the liner zone between 20°C and 180°C, this sensor has a sensitivity of 0.004°C^{-1}.

7. CONCLUSION

The response to temperature variations on the bending light power loss of a multimode optical fiber with different bent fiber curvature radii have been analyzed. It has been found that the bending losses due to the internal optical fiber numerical aperture variations increase when the fiber bending angle increase. The more important losses are caused by refraction and tunnel effects. We have shown that a bent optical fiber can be used as a temperature transducer. The use of an optical fiber curved during its manufacturing at high temperature allows to minimize some residual mechanical effects, and allows to use rigorously the geometrical approach to describe the light propagation, to evaluate the losses in light power output values and to calibrate the temperature sensor. We have shown that if we use, for example a silica/ silicone fiber as a transducer, we can obtain good performances with an excellent sensitivity and an excellent linearity associated to a large temperature measurement range, mainly because the thermo-optic effect value on the silicone is negative and important.

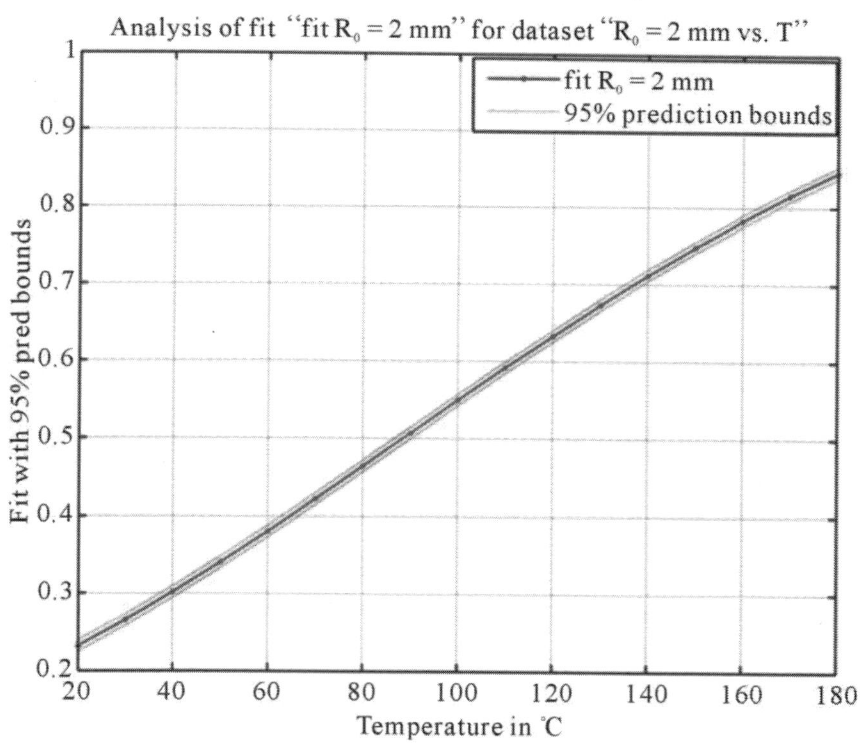

Figure 9. The temperature response curve fitting for R_0 = 2 mm between 20°C and 180°C.

REFERENCES

1. M. Remouche, R. Mokdad and A. Chakari, "Intrinsic Integrated Optical Temperature Sensor Based on Waveguide Bend Loss," Optics & Laser Technology, Vol. 39, No. 7, 2007, pp. 1454-1460.

2. M. Remouche, R. Mokdad and M. Lahrashe, "Intrinsic Optical Fiber Temperature Sensor Operating by Modulation of the Local Numerical Aperture," Optical Engineering, Vol. 46, No. 2, 2007, p. 024401.

3. A. W. Snyder and J. D. Love, "Optical Waveguide Theory," Chapman & Hall, Upper Saddle River, 1983.

4. A. Zendehnam, M. Mirzaei1 and A. Farashiani, "Investigation of Bending Loss in a Single-Mode Optical Fibre," Pramana Journal of Physics-Indian Academy of Sciences, Vol. 74, No. 4, 2010, pp. 591-603.

5. R. Ulrich, S. C. Rashleigh and W. Eickhoff, "BendingInduced Birefringence in Single-Mode Fibers," Optics Letters, Vol. 5, No. 6, 1980, pp. 273-275.

6. E. A. J. Marcatilli, "Bends in Optical Dielectric Guides," Bell System Technical Journal, Vol. 48, No. 7, 1969, pp. 2103-2132.

7. D. Gloge, "Bending Loss in Multimodes Fibers with Graded and Upgraded Core Index," Applied Optics, Vol. 11, No. 11, 1972, pp. 2187-2506.

8. D. Marcuse, "Influence of Curvature on the Losses of Doubly Clad Fibers," Applied Optics, Vol. 21, No. 23, 1982, pp. 4208-4213.

9. M. Semenkoff, "Bending Effect on Light Propagation in an Optical Fiber: Application to a Temperature Sensor," Optics and Lasers in Engineering, Vol. 17, No. 3-5, 1992, pp. 179-186

10. D. Marcuse, "Bending Losses of the Asymmetric Slab Waveguide," Bell System Technical Journal, Vol. 50, No. 8, 1969, pp. 2551-2563.

11. A. W. Snyder and D. J. Mitchell, "Generalized Fresnel's Laws for Determining Radiation Loss from Optical Waveguides and Curved Dielectric Structures," Optik, Vol. 40, No. 4, 1974, pp. 438-459.

12. A. W. Snyder and D. J. Mitchell, "Bending Losses of Multimode Optical Fibers," Electron Letter, Vol. 10, No. 1, 1974, pp. 11-13.

13. J. Dakin, "Optical Fiber Sensors," Artech House, Norwood, 1988.

14. T. S. Yu Francis, "Fiber Optical Sensors," Marcel Dekker, Inc., New York, 2002.

15. F. Pang; W. Liang and W. Xiang, "Temperature-Insensitivity Bending Sensor Based on Cladding-Mode Resonance of Special Optical Fiber," IEEE Photonics Technology Letters, Vol. 21, No. 2, 2009, pp. 76-78.

16. R. P. Hu and X. G. Huang, "A Simple Fiber-Optic Flowmeter Based on Bending Loss," IEEE Sensors Journal, Vol. 9, No. 12, 2009, pp. 1952-1955.

17. R. M. Gavalis, P. Y. Wong and J. A. Eisenstein, "Localized Active-Cladding Optical Fiber Bend Sensor," Optical Engineering, Vol. 49, No. 6, 2010, p. 064401.

18. X. Chen, C. Zhang and D. J. Webb, "Bragg Grating in a Polymer Optical Fibre for Strain, Bend and Temperature Sensing," Measurement Science and Technology, Vol. 21, No. 9, 2010, p. 094005.

19. J. H. Kuang, P. C. Chen and Y. C. Chen, "Plastic Optical Fiber Displacement Sensor Based on Dual Cycling Bending," Sensors, Vol. 10, 2010, pp. 10198-10210.

20. J. Zhang, H. Liu and X. Wu, "Curvature Optical Fiber Sensor by Using Bend Enhanced Method," Frontiers of Optoelectronics in China, Vol. 2, No. 2, 2009, pp. 204- 209.

21. N. singgh, V. Mishra and S. C. Jain, "Enhanced Sensitivity Refractive Index Sensor Based on Segmented Fiber with Bending," Indian Journal of Pure & Applied Physics, Vol. 47, No. 9, 2009, pp. 655-657.

22. Z. Zhang, G. Z. Xiao and P. Zhao, "Planar Wave GuideBased Silica-Polymer Hybrid Variable Optical Attenuator and Its Associated Polymers," Applied Optics, Vol. 44, No. 20, 2005, pp. 2402-2408.

23. H. Ma, A. K.-Y. Jen and L. R. Dalton, "Polymer-Based Optical Waveguides: Material, Processing, and Devices," Adv Mater, Vol. 14, No. 19, 2002, pp. 1339-1365.

24. M. Zhou, "Low-Loss Polymeric Material for Waveguide Components in Fiber Optical Telecommunication," Optical Engineering, Vol. 41, No. 7, 2002, pp. 1631-1643.

25. L. Eldada, "Advance in Polymer Integrated Optics," IEEE Journal of Selected Topics in Quantum Electronics, Vol. 6, No. 1, 2000, pp. 54-68.

26. T. Watanabe, N. Ooba and S. Hayashida, "Polymeric Optical Waveguide Circuits Formed Using Silicone Resin," Journal of Lightwave Technology, Vol. 16, No. 6, 1998, pp. 1049-1055.

6

NANO-PLASMONIC FILTERS BASED ON TOOTH-SHAPED WAVEGUIDE STRUCTURES

Xu Guang Huang[1] and Jin Tao[1]

[1] Key Laboratory of Photonic Information Technology of Guangdong Higher Education Institutes, South China Normal University, Guangzhou, China

1. INTRODUCTION

Along with development of human society and technology, it becomes more dependable on the miniaturization and integration of semiconductor components, circuits and devices. The performance of integrated circuits, such as micro-processor, is in accordance with the famous Moore's law that the number of transistors placed inexpensively on an integrated circuit doubles approximately every two years. However, the integration of modern electronic components and devices for information communication and processing have been approaching its fundamental speed and bandwidth limitation, because the ultra-intensive electrical interconnects have an increased effective resistor-capacitor (RC) time constant that increases the time of charging and discharging [1, 2]. This has caused an increasing serious problem that hinders further development in many fields of modern science and technology. Using light signals instead of electronic is one of the most promising solutions. The speed of optical signal is on the order of 10^8 m/s, which is about 3 orders of the saturation velocity of electrons in a semiconductor such as silicon [3]. However, a major problem with using light as information carrier in conventional optical devices is the poor performance of integration and miniaturization. Dielectric waveguides are basic components and cannot allow the localization of electromagnetic waves into subwavelength-scale regions because of diffraction limit $\lambda_0/2n$, here λ_0 is the wavelength of the light in the free space and n is the refractive index of the dielectric. Photonic crystal (PC) structures and devices have been studied by many researchers since E. Yablonovitch and S. John ' s two milestone published papers in 1987 [4, 5], which confirmed that the light can be confined in the nanoscale. However, the dimensions of the PC system are on the order of the wavelength or even larger, making them less appropriate for nano-scale optical elements integration.

Surface Plasmon polaritons (SPPs) are electromagnetic waves that propagate along the interface of metal and dielectric. In recent years, plasmonics is called the area of nanophotonics under the light diffraction limit that studies the transmission characteristics, localization, guiding of the SPP mode using metallic nanosturctures. The plasmonic waveguides are the most basic components and have been given much attention. Various kinds of metallic nanostructures have been proposed for SPP guiding. Generally these structures could be classified into three big categories: 1) chains of metal nanoparticles [6] and cylindrical metallic nanorods with various geometries [7]. 2) Metal-dielectirc-metal (MDM) or Metal-insulator-metal (MIM) plasmonic waveguides, including groove channel structures in metallic substrates [8], slot waveguide [9]. 3) Dielectric-metal-dielectric (DMD)/ Insulator-metal-Insulator (IMI) waveguide [10]. It should be noted that not all these plasmonic structures can be used for guiding SPP mode to achieve subwavelength localization. Among them, MDM/MIM plasmonic waveguide can propagate SPP mode in the subwavelength scale with relatively low dissipation and large propagation distance. Our following proposed structures are mainly based on the MDM/MIM structures.

Wavelength selection is one of key technologies in fields of optical communication and computing. To achieve wavelength filtering characteristics, plasmonic Bragg reflectors and nanocavities have been proposed. They include the metal hetero-structures constructed with several periodic slots vertically along a metal-dielectric-metal (MDM) waveguide [11], the Bragg grating fabricated by periodic modulating the thickness of thin metal stripes embedded in an insulator [12] and the periodic structure formed by changing alternately two kinds of the insulators [13,14]. Lately, a high-order plasmonic Bragg reflector with a periodic modulation of the core index of the insulators [15], and a structure with periodic variation of the width of the dielectric in MDM waveguide [16] have been proposed. Most of the periodic structures mentioned above, however, have total length of micrometers and relatively high propagation loss of several decibels.

In this chapter, we present our recent work on compact nano-plasmonic waveguide filters based on T-series and nano-capillary structures. In section 2, we introduce the novel nanometeric plasmonic filter in a tooth-shaped MIM waveguide and give an analytic model based on the scattering matrix method. In section 3, we investigate the characteristics of double-side teeth-shaped nano-plasmonic waveguide. In section 4 and section 5, we introduce the multiple multiple-teeth-shaped plasmonic filters and asymmetrical multiple teeth-shaped narrow pass band subwavelength filter. In section 6, we introduce a wavelength demultiplexing structure based on metal-dielectric-metal plasmonic nano-capillary resonators. Finally, we make a conclusion.

2. SINGLE-TOOTH SHAPED PLASMONIC WAVEGUIDE FILTER [17]

To begin with the dispersion relation of the fundamental TM mode in a MIM waveguide (shown in the inset of Fig. 1) is given by [18]:

$$\varepsilon_{in} k_{z2} + \varepsilon_m k_{z1} \coth\left(-\frac{ik_{z1}}{2} w\right) = 0, \tag{1}$$

with k_{z1} and k_{z2} defined by momentum conservations:

$$k_{z1}^2 = \varepsilon_{in} k_0^2 - \beta^2, \quad k_{z2}^2 = \varepsilon_m k_0^2 - \beta^2 . \tag{2}$$

Where εin and εm are respectively dielectric constants of the insulator and the metal, $k0=2\pi/\lambda 0$ is the free-space wave vector. The propagation constant β is represented as the effective index $neff = \beta/k0$ of the waveguide for SPPs. The real part of *neff* of the slit waveguide as a function of the slit width at different wavelengths is shown in Fig. 1. It should be noted that the dependence of *neff* on waveguide width is also suitable to the region of the tooth waveguide with the tooth width of *wt* shown in Fig. 2. The imaginary part of *neff* is referred to the propagation length which is defined as the length over which the power carried by the wave decays to 1/e of its initial value: $Lspps=\lambda 0/[4\pi\ \mathrm{Im}(neff)]$. In the calculation above and the following simulations, the insulator in all of the structures is assumed to be air (εin=1), and the frequency-dependent complex relative permittivity of silver is characterized by

Drude model: $\varepsilon_m(\omega) = \varepsilon_\infty - \omega_p^2 / \omega(\omega + i\gamma)$. Here $\omega_p = 1.38 \times 10^{16} Hz$ is the bulk plasma frequency, which represents the natural frequency of the oscillations of free conduction electrons. $\gamma = 2.73 \times 10^{13} Hz$ is the damping frequency of the oscillations, ω is the angular frequency of the incident electromagnetic radiation, ε∞ stands for the dielectric constant at infinite angular frequency with the value of 3.7 [16]

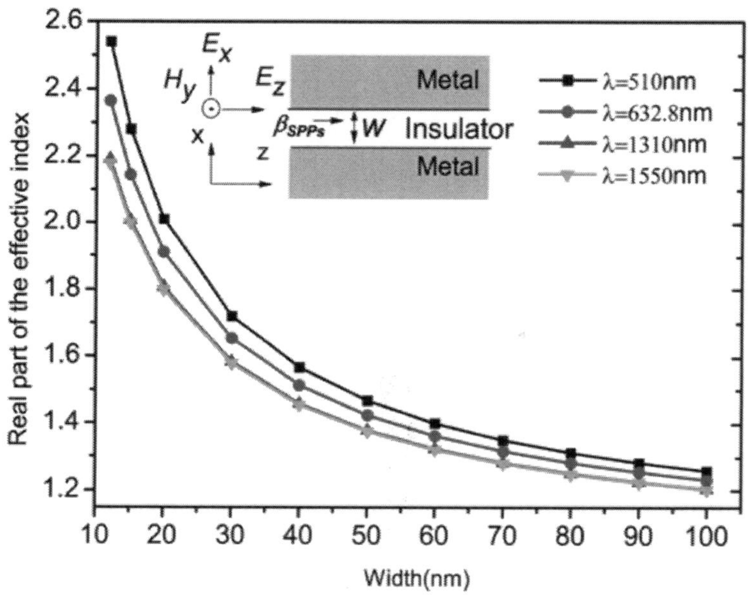

Figure 1. Real part of the effective of refraction index versus the width of a MIM slit waveguide structure.

The tooth-shaped waveguide filter is shown in Fig. 2. In the following FDTD simulations, the grid sizes in the *x* and *z* directions are chosen to be 5nm×5nm. Since the width of the waveguide is much smaller than the operating wavelength in the structure, only fundamental waveguide mode is supported. Two power monitors are respectively set at the positions of *P* and *Q* to detect the incident and transmission fields for calculating the incident power of *Pin* and the transmitted power of *Pout*. The transmittance is defined to be *T=Pout/Pin*. The length of *L* is fixed to be 300nm, while the tooth width and depth are respectively *wt*=50nm and *d*=100nm. The tabulation of the optical constants of silver [19] is used in the simulation. As shown in Fig. 3(a), the tooth-shaped waveguide is of a filtering function: A transmission dip occurs at the free space wavelength nearly 784nm with the transmittance of ~0%. The maximum transmittance at the wavelengths longer than 1700nm is over 90%. The contour profiles of the field distributions around the tooth-shaped area at different wavelengths are shown in Figs. 3(b)-3(d). The filtering structure is distinguished from the Bragg reflectors based on periodical heterostructure.

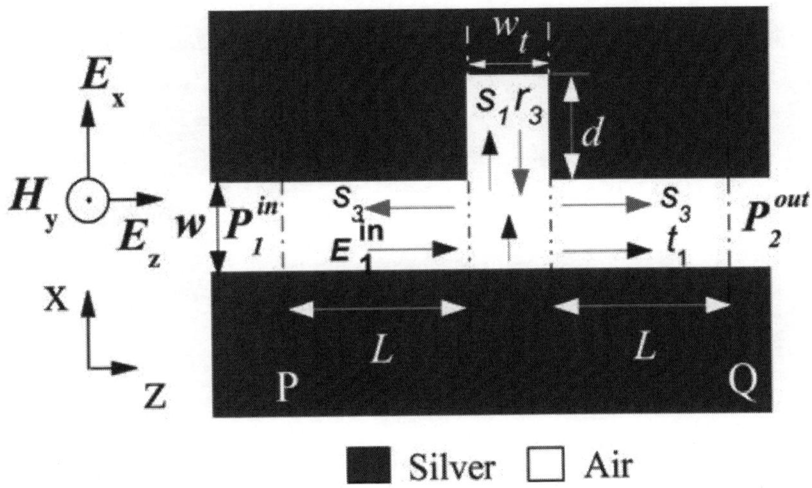

Figure 2. The structure schematic of a single tooth-shaped waveguide filter, with the slit width of *w*, the tooth width of *wt*, and the tooth depth of *d*.

The phenomenon above can be physically explained in the scattering matrix theory [20] as follows:

$$\begin{pmatrix} E_1^{\text{out}} \\ E_2^{\text{out}} \\ E_3^{\text{out}} \end{pmatrix} = S \cdot \begin{pmatrix} E_1^{\text{in}} \\ E_2^{\text{in}} \\ E_3^{\text{in}} \end{pmatrix}, \tag{3}$$

Where $S = \begin{bmatrix} r_1 & t_1 & s_3 \\ t_1 & r_1 & s_3 \\ s_1 & s_1 & r_3 \end{bmatrix}$, r_i, t_i and *si* (*i*=1,2,3) are respectively the reflection, transmission and splitting coefficients of a incident beam from Port *i*

(i=1,2,3), caused by the structure. Eini and Eouti stand for the fields of incident and output beams at Port i, respectively. Using the fact that |S|=1, one can obtain:

$$r_1^2 r_3 + 2t_1 s_1 s_3 - 2r_1 s_1 s_3 - t_1^2 r_3 = 1. \tag{4}$$

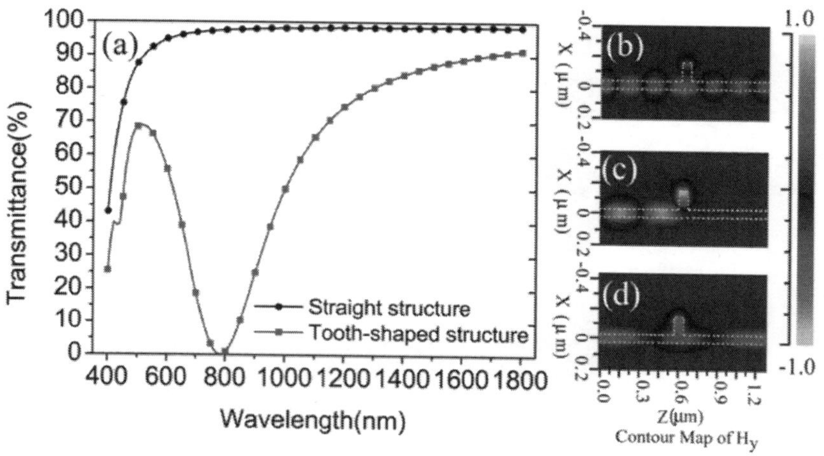

Figure 3. a) Transmission of the single tooth-shaped MIM waveguide compared with a straight MIM slit waveguide. The width of the waveguide is w=50nm, and the tooth width and depth are respectivelywt=50nm and d=100nm. The contour profiles of field Hy of the tooth-shaped waveguide at different wavelengths of (b) λ=510nm, (c) λ=783nm, and (d) λ=1550nm.

For the case of E_2^{in} =0, one has

$$E_2^{out} = t_1 E_1^{in} + s_3 E_3^{in}, \tag{5}$$

in which E_3^{in} is given as follows:

$$E_3^{in} = s_1 E_1^{in} \exp(i\phi(\lambda))(1 + r_3 \exp(i\phi(\lambda)) + r_3^2 \exp(2i\phi(\lambda)) + ...) = \frac{s_1 E_1^{in}}{1 - r_3 \exp(i\phi(\lambda))} \exp(i\phi(\lambda)), \tag{6}$$

where the phase delay $\phi(\lambda)=\frac{4\pi}{\lambda}n_{eff}\cdot d+\Delta\varphi(\lambda)$, and $\Delta\varphi(\lambda)$ is the phase-shift caused by the reflection on the air-silver surface. Combined Eq. (5) and Eq. (6), the output field at Port 2 is derived as:

$$E_2^{\text{out}}=t_1E_1^{\text{in}}+\frac{s_1s_3E_1^{\text{in}}}{1-r_3\exp(i\phi(\lambda))}\exp(i\phi(\lambda)), \tag{7}$$

Therefore, the transmittance *T* from Port 1 to Port 2 is given by:

$$T=\left|\frac{E_2^{\text{out}}}{E_1^{\text{in}}}\right|^2=\left|t_1+\frac{s_1s_3}{1-r_3\exp(i\phi(\lambda))}\exp(i\phi(\lambda))\right|^2. \tag{8}$$

It can be seen from Eq. (8) that, if the phase satisfies $\phi(\lambda)=(2m+1)\pi$ (m=0,1,2...), the two terms inside the absolute value sign on the right of the equation will cancel each other (as it can be seen inFig. 3(c)), so that the transmittance *T* will become minimum. Therefore, the wavelength λm of the transmission dip is determined as follows:

$$\lambda_m=\frac{4\cdot n_{eff}\cdot d}{(2m+1)-\frac{\Delta\varphi(\lambda)}{\pi}}. \tag{9}$$

It can be seen that the wavelength λm is linear to the tooth depth *d*, and depends on tooth width *wt*, through the somewhat inverse-proportion-like relationship between *neff* and *wt* shown in Fig. 1.

Figure 4(a) shows the transmission spectra of the waveguide filters with various tooth widths of *wt*. The maximum transmittance can reach 97%. Figure 4(b) shows the wavelength of the trough vs. the tooth width of *wt*. The primary dip of the transmission moves very significantly to short wavelength (blue-shift) with the increase of *wt* for *wt*<20nm. The shift rate rapidly becomes small after *wt*>20nm, and tends to be saturated when *wt*>200nm. As revealed in the Eq. (9), the above relationship between the dip position and *wt* mainly results from the contribution of the inverse-proportion-like dependence of *neff* on *wt*. The change rate of Δ*neff* /Δ*wt* within the tooth width of 20nm is much higher than that of Δ*neff* /Δ*wt* after *wt*>20nm, as shown in Fig. 1, and becomes finally saturated after *wt*>200nm. Obviously, tooth width *wt* should be chosen within the range of 20-200nm to avoid the critical behavior and the difficulty in fabrication process.

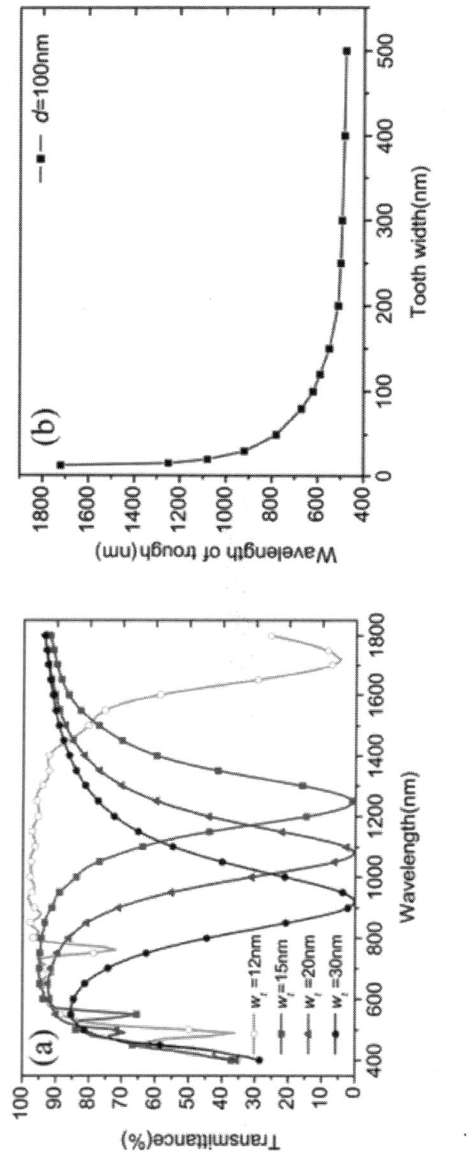

Figure 4. a) Transmission spectra of the waveguide filters with various tooth widths of *wt*, at a fixed tooth depth of *d*=100nm and the slit width of *w*=50nm. (b) The wavelength of the trough versus the tooth width of *wt*.

Figure 5(a) shows the transmission spectra of the filters with different tooth depths of d. It is found that the wavelength of the transmission dip shifts to long wavelength with the increasing of d. Figure 5(b)reveals that the wavelength of the transmission dip has a linear relationship with the tooth depth, as our expectation in Eq. (9). Therefore, one can realize the filter function in various required wavelength with high performance by changing the width or/and the depth of the tooth. For example, to obtain a filter with a transmission dip at the wavelength of 1550nm, the structural parameters of $wt=w=50$nm and $d=237.5$nm can be chosen.

3. DOUBLE-SIDE TEETH-SHAPED NANO-PLASMONIC WAVEGUIDE FILTERS [21]

Figure 6(a) shows the structure of a proposed double-side teeth-shaped waveguide filter. The waveguide width w and the distance L are fixed to be 50nm and 300 nm. d_1 and d_2 are the depths of the teeth on each side of a MIM waveguide. The transmission spectra of the double-side teeth-shaped waveguide filter and the single-side tooth-shaped waveguide filter are shown in Fig. 6(b), which is obtained with the FDTD method. One can see that there is one dip at the free-space wavelength of nearly 1550nm with the transmittance of ~0%. An analytic model to explain the filtering function of single-side tooth-shaped waveguide structure based on multiple-beam-interference and scattering matrix has been given in the section 1

It can be seen that the wavelength λ_m is linear to the tooth depth d and depends on tooth width $w$through the somewhat inverse-proportionlike relationship between n_{eff} and w shown in Fig. 1. The dip width of the double-side structure is wider compared with that of the single-side tooth-shaped structure. For better understanding the characteristics of double-side teeth structure, we next discuss the dependence of its filtering spectrum on the symmetry of teeth depth.

Figure 7 shows the transmission spectra of the double-side teeth-shaped waveguide filters with different teeth depths of d_1 and d_2. From Fig. 7(a), one can see that there are two transmission dips at the free-space wavelengths of nearly 790nm and 1230nm with $d_1=100$nm and $d_2=180$nm. It is found that the wavelength of the second dip shifts to a long wavelength with the increasing of d_2 while the first trough keeps unchanged due to a fixed value of $d_1=100$nm. From Fig. 7(b), we can see that the first dip shifts to a long wavelength with the increasing of d_1 while the position of the second dip is fixed due to a given d_2=180nm.

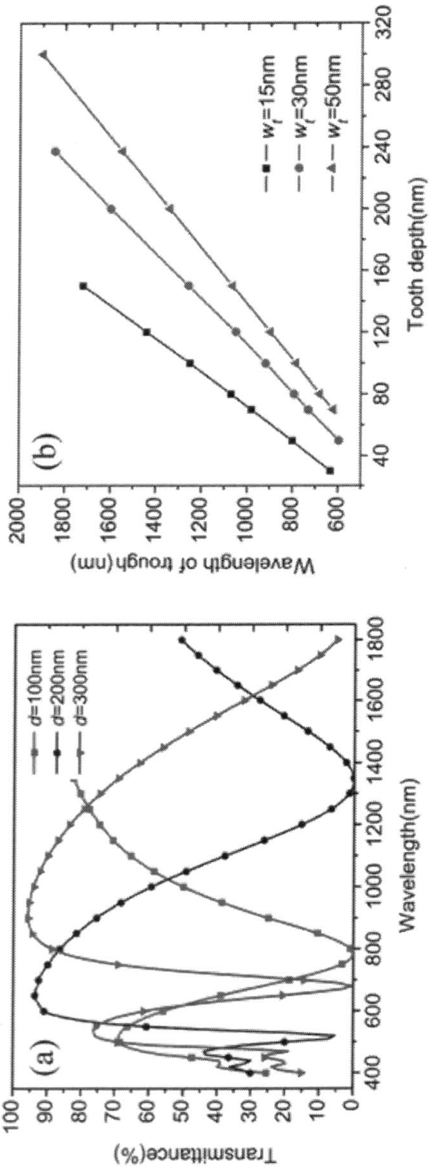

Figure 5. a) Transmission spectra of the waveguide filters with different tooth depths of *d*, and with a given tooth width of *wt*=50nm and the slit width of *w*=50nm. (b) The wavelength of the transmission dip vs. the tooth depth of *d* with *wt*=15nm, *wt*=30nm and *wt*=50nm.

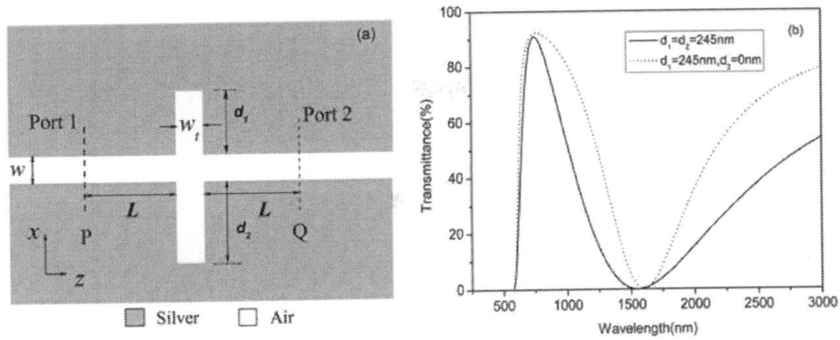

Figure 6. a) Schematic of a double-side teeth-shaped nano-plasmonic waveguide. The double-side teeth-shaped structure can be asymmetrical, if $d_1 d_2$. (b) The transmission spectra of a symmetrical double-side teeth-shaped waveguide filter with w_t=50nm, d_1=d_2=245nm and a single-side tooth-shaped waveguide filter with w_t=50nm, d_1= 245nm.

To better understand the origin of the two dips, the transmission spectra of an asymmetrical double-side teeth-shaped waveguide filter with d_1=100nm and d_2=180nm and two single tooth shaped structures with the tooth depths of 100nm and 180nm are all shown in Fig. 8. From it, one can see that the trough positions of the two single tooth waveguide filters with the tooth depths d_1=100nm andd_2=0 as well as d_1=0 and d_2=180nm overlap with the positions of the dips of the asymmetrical double-side teeth-shaped structure, which means the positions of the two dips of the double-side teeth-shaped structure are almost respectively determined by its two different single-tooth parts. Therefore, one can realize the filter function in various required wavelengths by respectively changing the depths of $d_1$and d_2 of the two single teeth.

It is also interest to address a staggered double-side teeth-shaped MIM waveguide structure (shown inFig. 9(a)). *Ls* stands for the shift length between two single teeth. A typical transmittance of the staggered double-side teeth-shaped waveguide filter with *wt*=50nm, d_1=d_2=260.5nm, and *Ls* =250nm is shown in Fig. 9(b). A wide bandgap occurs around =1700nm with the bandgap width (defined as the difference between the two wavelengths at of which the transmittance is equal to 1%) is 660nm, and the transmittance of passband is over 85%. The filter's feature can be attributed to the interference superposition of the reflected and transmitted fields from each tooth of the double-side structure.

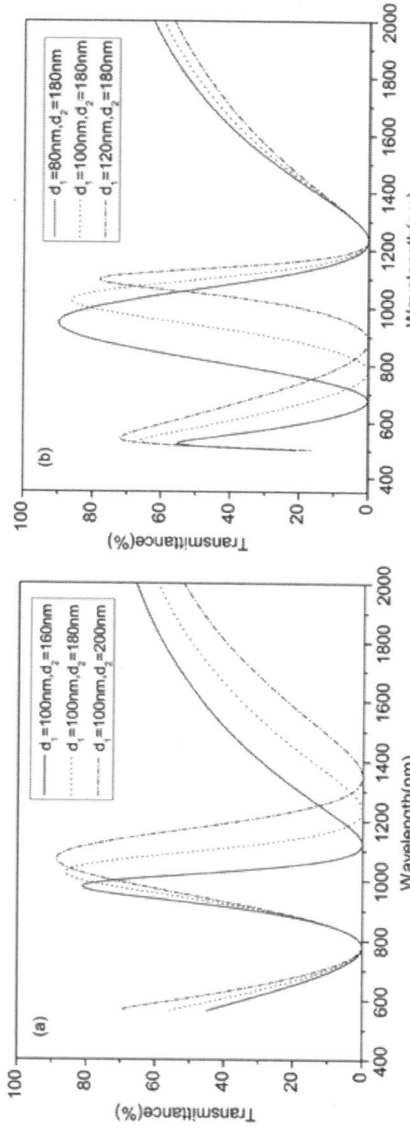

Figure 7. a)Transmission spectra of the asymmetrical double-side teeth-shaped waveguide filters for different tooth depths of d_2 with a fixed d_1=100nm and w_t=50nm. (b) Transmission spectra of the asymmetrical double-side teeth-shaped waveguide filters for different tooth depths of d_1 with a fixedd_2=180nm and w_t=50nm.

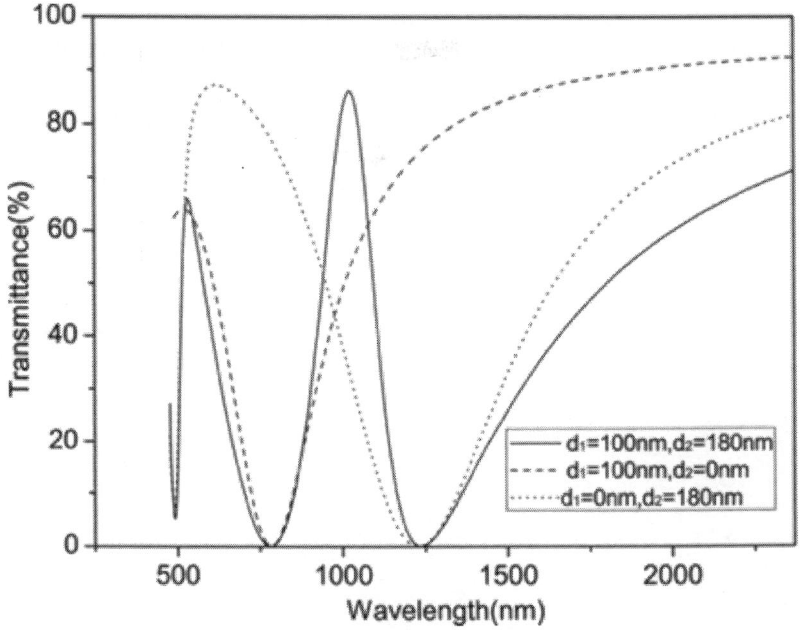

Figure 8. Transmission spectra of the two single-tooth waveguide filters and an asymmetrical double-side teeth-shaped structure with a given tooth width of *wt*=50nm and a slit width of *w*=50nm.

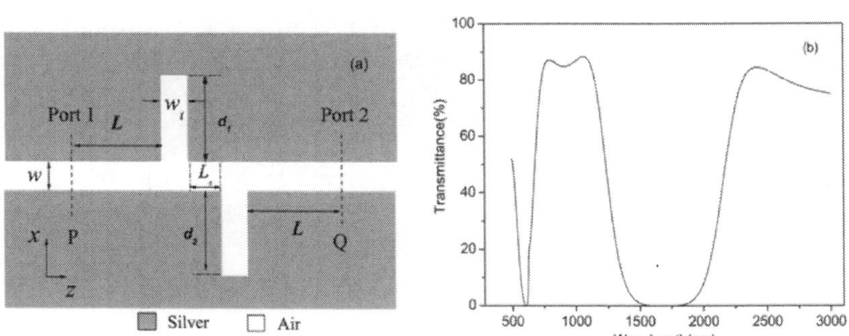

Figure 9. a) Schematic of a staggered double-side teeth-shaped nano-plasmonic waveguide with a shift length of L*s*. (b) The transmittance of the staggered double-side teeth-shaped waveguide filter with w_t=50nm, d_1=d_2=260.5nm, L*s*=250nm.

Figure 10 shows the central wavelength of the bandgap as a function of the double-side teeth depth of d. During the simulation, we set $d=d_1=d_2$. The FDTD simulation results reveal that the relationship between the central wavelength of the bandgap and the double-side teeth depth of d is a linear function. It reveals that the central wavelength of the bandgap shifts toward long wavelength with the increasing of the teeth depth of d.

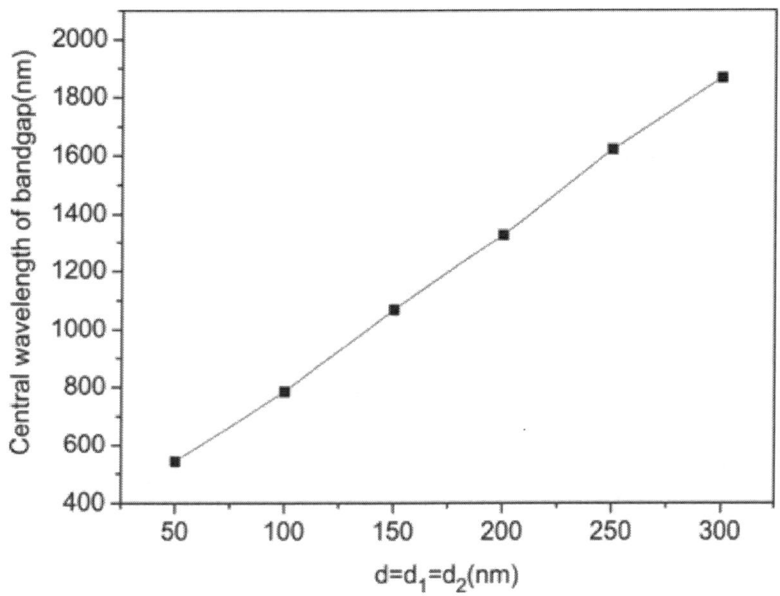

Figure 10. The central wavelength of the bandgap as a function of the double-side teeth depth of d at teeth width of 50nm.

Figure 11 shows how the shift length of the teeth influences or modifies the filtering spectrum of the structure. One can see that central bandgap shifts left and becomes wider with increasing the shift length of L*s*. It reveals the filtering characteristics of the structure depend on the phase difference between the plasmon waves passing through the tooth. The optimized filtering response with a sharp left band-edge and high passband-transmittances over 85% can be achieved when the shift length is around 200nm.

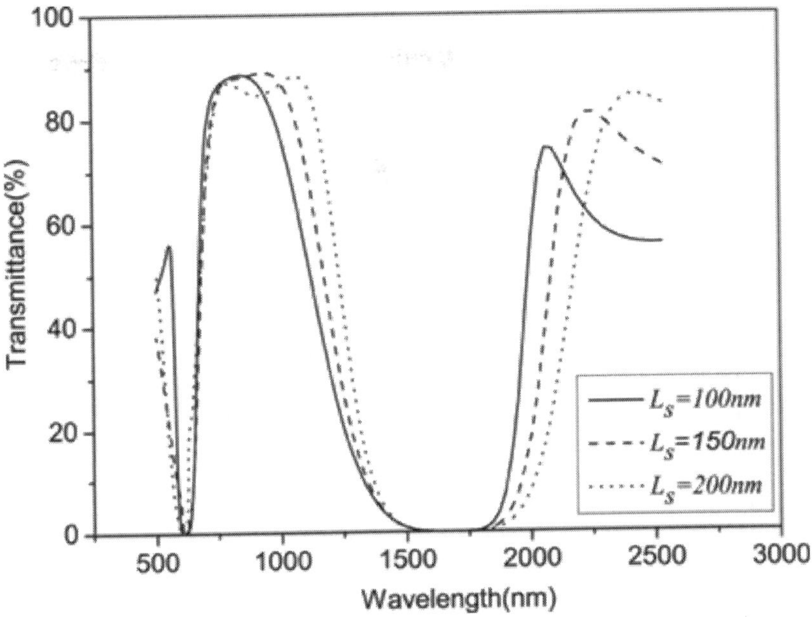

Figure 11. The transmittance of the two-sided staggered teeth-shaped waveguide filter for different shift lengths with w_t=50nm, d_1=d_2=260.5nm.

4. A MULTIPLE-TEETH-SHAPED WAVEGUIDE BANDGAP FILTER [22]

It is straight forward and basic interest to expand a single tooth structure to multiple-teeth structure (shown in Fig. 12(a)), and check the difference between them. For the sake of comparison, the waveguide width w and the distance L are fixed to be 50nm and 300nm. Λ and N are the period and the number of rectangular teeth, respectively. w*gap* stands for the width of the gap that between any two adjacent teeth, and one has wt+wgap=Λ. A typical transmittance of the multiple-teeth-shaped waveguide filter with *wt*=50nm, Λ=150nm, d=260.5nm and N=5 is shown in the Fig. 12(b), which is obtained with FDTD method. A wide bandgap occurs around λ =1.55μm with the bandgap width 590nm, and the transmittance of passband is over 90%. The filter's feature can be attributed to the superposition of the reflected and transmitted fields from each of the five single tooth-shaped components. Figure 13(a) shows the central wavelength of the transmittance bandgap, while the right y-axis displays the bandgap width of the waveguide filter as a function of teeth depth d at various *wt*with the same Λ=150nm and N=5. The FDTD simulation results reveal that the relationship between the central wavelength of the bandgap and the teeth depth d is a linear function for any *wt*, which is indeed the one of expectations in Eq. (9). Figure 13(b) shows the central wavelength of the bandgap and the bandgap width as a function of teeth width of *wt* at various teeth depths. As revealed in the Eq. (9), the relationship between the bandgap position and *wt* mainly results from the contribution of the inverse-proportion-like dependence of n*eff* on *wt* as shown in Fig. 1(a). Obviously, teeth width *wt*should be chosen within the range of 50-100nm with a slope of dλcenter/dwt≈5d to avoid a large value of dλcenter/dwt≈20 when *wt* <45nm in Fig. 13(b), and to reduce the sensitivity of the central wavelength of bandgap in fabrication process. Therefore, one can realize the filter function at various required wavelengths with high performance, by choosing the width or/and the depth of the teeth.

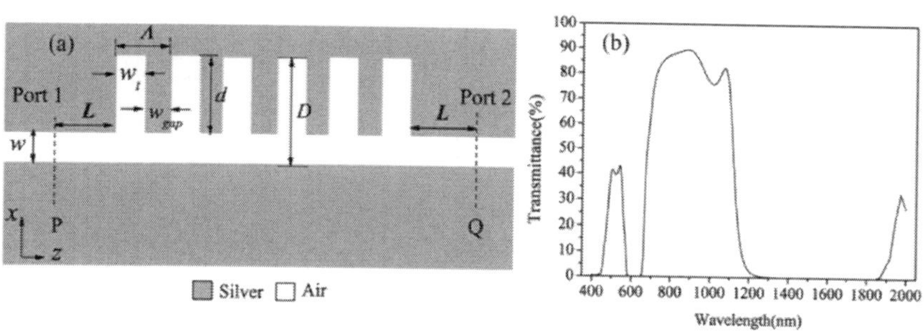

Figure 12. a) Schematic of a multiple-teeth-shaped MIM waveguide structure. (b) The transmittance of the multiple-teeth-shaped waveguide filter with *wt*=50nm, Λ=150nm, d=260.5nm and N=5.

For the multiple-teeth-shaped structure with the parameters of $n_{eff,teeth} = 1.070$ for the width of D=d+w=(260.5+50)nm, neff,wg=1.375 for w=50nm, and the teeth period of Λ=wt+wgap=150nm in z-axis direction, one can see $w_t \mathrm{Re}(n_{eff,teeth}) + w_{gap} \mathrm{Re}(n_{eff,wg}) \approx$. Thus the structure does not follow the Bragg condition in z-axis direction.

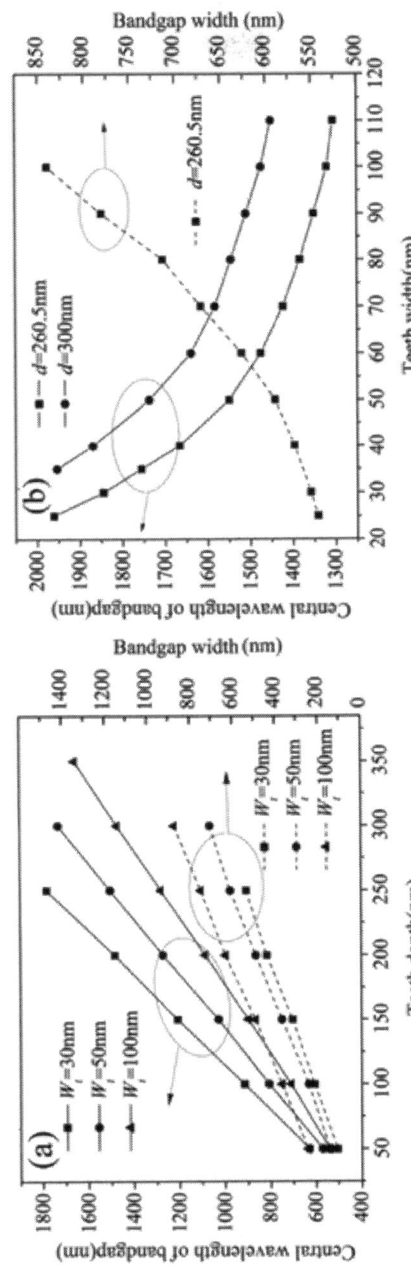

Figure 13. a) The central wavelength of the bandgap and the bandgap width as a function of the teeth depth of d at various teeth widths. (b) The central wavelength of the bandgap and the bandgap width as a function of teeth widths of *wt* at various teeth depths.

Figure 14(a) and (b) show the transmission spectra of a multiple-teeth-shaped waveguide filter at different periods Λ and period numbers N. As one can see from Fig. 14(a), when Λ=100nm is chosen, the coupling of the SPPs waves between two adjacent teeth is strong which causes the central bandgap wavelength to shift left and the bandgap to be wider. When the period equals to Λ=200nm, the coupling between any two adjacent teeth becomes very weak. One can see in Fig. 14(b) that, the forbidden bandwidth increases little with the changing of the period number from N=3 to 7, while the transmittance of the passband decreases from 93% to 86%. The reason for the decreasing in transmittance can be attributed to the increasing of the propagation loss of the lengthened structure with a large period number. From the simulation results, a tradeoff period number N=4 is the optimized number with the transverse filter length of 4×150nm, which is ~5 times shorter than the previous grating-like filter structures.

5. A NARROW BAND SUBWAVELENGTH PLASMONIC WAVEGUIDE FILTER WITH ASYMMETRICAL MULTIPLE TEETH-SHAPED STRUCTURE [23]

The asymmetrical multiple-teeth-shaped structure is shown in Fig. 15(a), which is composed of two sets of multiple-teeth with two different teeth depths. The short set has three teeth, and the long set has four teeth. Λ, N_1 and N_2, are the period, the numbers of short rectangular teeth and long teeth, respectively. *wgap* stands for the width of the gap between any two adjacent teeth in multiple-teeth structure, and one has wt+wgap=Λ. The separation between the 3rd short tooth and 1st long tooth is *ws*. The length of L and the waveguide width w are, respectively, fixed to be 150nm and 50nm. In Fig. 15(a) we set d_1=148nm, d_2=340nm, *wt*=50nm, *wgap*= *ws* =84nm. Figure 15(b) shows a typical transmission spectrum of the asymmetrical multiple-teeth-shaped structure using FDTD method.

One can see the maximum transmittance at the wavelength of 1287nm is nearly 90%, and the full-width at half-maximum (FWHM) is nearly 70nm which is much smaller than the bandgap width of 1300nm. The FWHM of the asymmetrical multiple-teeth-shaped structure is also smaller than our previous coupler-type MIM optical filter [24].

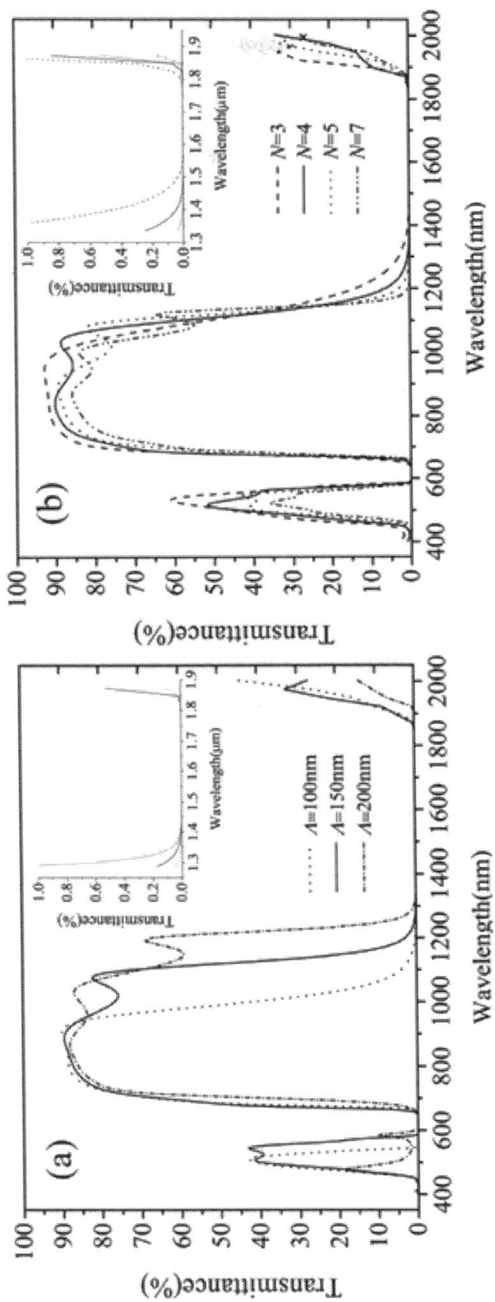

Figure 14. a) Transmittance spectra of multi-teeth filters with different periods and a fixed N=5, (b) Transmittance spectra of multi-teeth filters consisting of 3-7 periods with a fixed Λ=150nm.

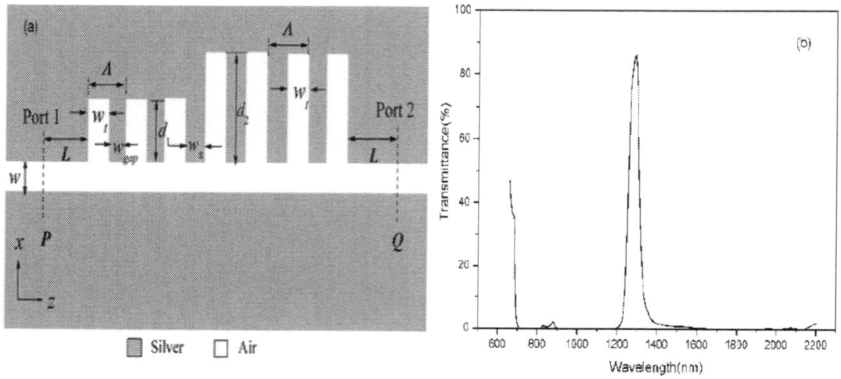

Figure 15. a) Schematic of an asymmetrical multiple-teeth structure consisted of two sets with different teeth depth. (b) The transmittance of the asymmetrical multiple-teeth-shaped waveguide filter with d_1=148nm, d_2=340nm. *wgap*= *ws* =84nm, N_1=3, and N_2=4.

In order to understand the origin of the narrow passband of the structure, the spectra of the transmission of a single-set of short three-teeth structure and a single-set of long four-teeth structure are calculated, and shown in Fig. 16. The parameters of the two structures are respectively equal to the parameters of the short teeth part and the long teeth part of the asymmetrical multiple-teeth-shaped structure (shown in Fig. 15(a)). One can see that the passband (or the bandgap) of the long teeth structure and the bandgap (or the passband) of the short teeth are overlapped from 800nm to 1200nm (or from 1450 to 1800nm), and then the transmission of the cascade of the two structure is very low within the two regions. Only the overlapping between the right edge of the passband of the long teeth structure and the left passband of the short teeth is non-zero. This is the reason why the wavelengths around 1300nm have a transmission peak in Fig. 15 (b).

Figure 17 shows the central wavelength of the narrow-band as a function of the variation of $\Delta d=\Delta d_1=\Delta d_2$. Δd is the increment of d_1 and d_2. The initial values of d_1 and d_2 are respectively, 128nm and 320nm. From the Fig. 17 can see that the central wavelength of the narrow-band linearly increases with the simultaneous increasing of d_1 and d_2. Figure 18 shows the dependence of transmission characteristic on separation *ws*. It is found that the transmission at the wavelength of 1287nm reaches the peak value when the separation of *ws* equals the gap of *wt*. Therefore, one can realize the narrow-bandwidth filter function at different required wavelengths by means of properly choosing the parameters of the device, such as the teeth-depth, the period or the separation of ws.

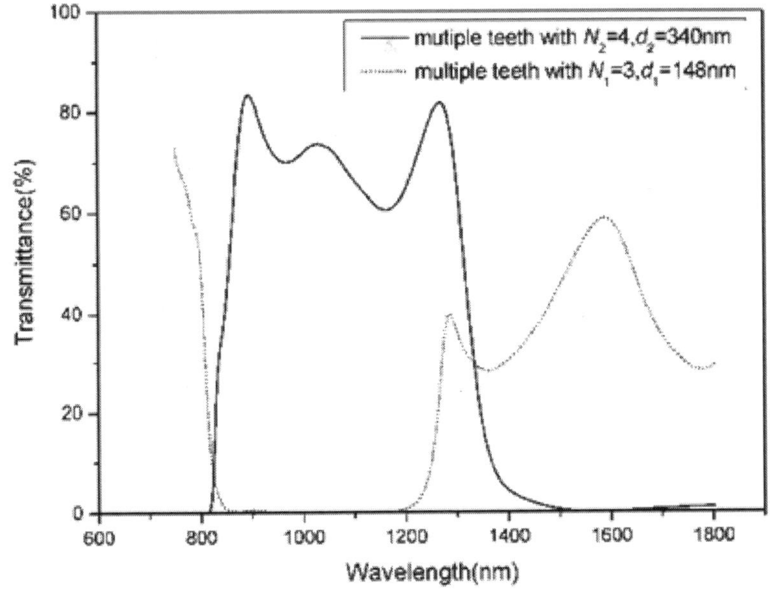

Figure 16. The transmission spectra of the single-set of multiple-teeth structure with d_1=148nm, N_1=3 and the single set of multiple-teeth structure with d_2=340nm, N_2=4, respectively.

6. A WAVELENGTH DEMULTIPLEXING STRUCTURE BASED ON METAL-DIELECTRIC-METAL PLASMONIC NANO-CAPILLARY RESONATORS [25]

The inset of Fig. 19 shows the nano-capillary resonators composed of two parallel metal plates with a dielectric core. Obviously, the structure can be treated as two MDM waveguides with different widths. Because the width of the lower MDM waveguide is much smaller than that of the upper part, here we call the lower (narrower) part as a nano-capillary. When the gap width w of the MDM waveguide is reduced below the diffraction limit, only a single propagation mode TM_0 can exist. The dielectric in the core of the structure is assumed to be air with a permittivity εd =1.

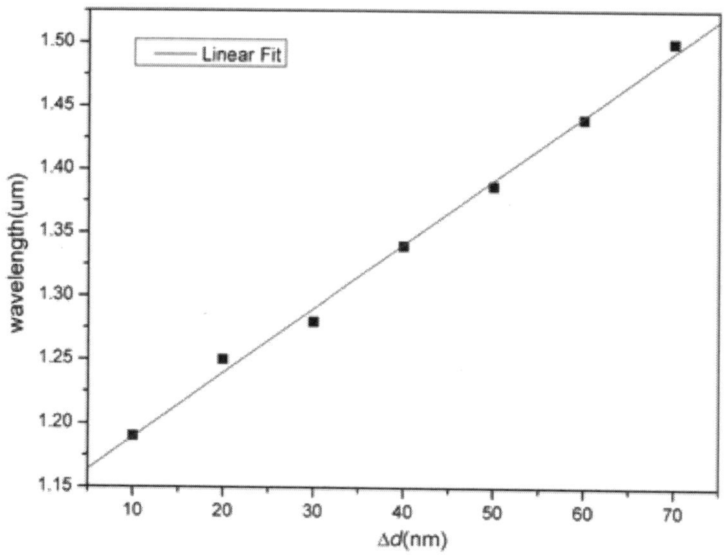

Figure 17. Central wavelength of the narrow-band as a function of the variation of $\Delta d=\Delta d_1=\Delta d_2$, $\Delta d_1$and $\Delta d_2$are respectively the increment of d_1 and d_2.

To fully understand how the width of the nano-capillary structure influences the SPPs propagation, the dependences of the effective index of SPPs on the width w at various wavelengths of the incident light are calculated and shown in Fig. 19. From the figure 19, one can see that the effective index of the waveguide decreases with increasing of w at the same wavelength. The effective index at short wavelength is larger than that at long wavelength, for a given width w. The effective index n_{eff2} of the nano-capillary can be larger than n_{eff3} of upper MDM part and n_1 of air. As shown in inset of Fig. 19, the waves will flow into the nano-capillary due to its higher effective index, when SPP waves propagate along the interface between metal and air. The wave transmitted into the capillary will be partly reflected at two ends of nano-capillary, because of index differences between n_{eff2} and n_{eff3} as well as n_1. One can expect the nano-capillary operates as a resonator. Resonance waves can be formed only in some appropriate conditions within nano-capillary segment. Defining $\Delta\phi$ to be the phase delay per round-trip in the nano-capillary, one has $\Delta\phi = 4\pi n_{eff} d / \lambda + \phi_r$, where $\phi_r \equiv \phi_1 + \phi_2$, ϕ_1 and ϕ_2 are respectively the phase shifts of a beam reflected on the entrance of the capillary and the junction connecting the nano-capillary and the upper MDM waveguide, and d is the length of the capillary. The waves propagating through the structure will be trapped within the nano-capillary when the following resonant condition is satisfied: $\Delta\phi = m \cdot 2\pi$.

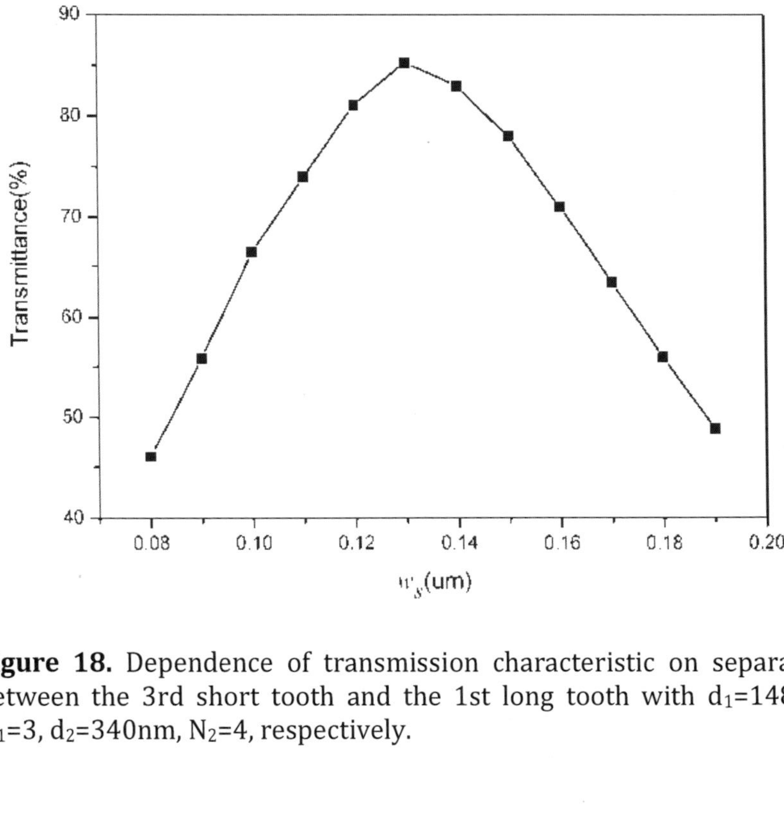

Figure 18. Dependence of transmission characteristic on separation between the 3rd short tooth and the 1st long tooth with d_1=148nm, N_1=3, d_2=340nm, N_2=4, respectively.

Here, positive integer m is the number of antinodes of the standing SPP wave. The resonant wavelengths can be obtained as follows:

$$\lambda_m = 2n_{eff}d/(m - \phi_r/\pi). \tag{10}$$

It can be seen that the wavelength λm is linear to the length and the effective index of the nano-capillary, respectively. Obviously, only the waves with the wavelength λm can stably exist in the nano-capillary, and thus partly transmit or drop into the output end of the nano-capillary. When wideband SPP waves incident into the structure, only the resonance waves with the wavelength λm can be selected and dropped by the nano-capillary. In other words, a transmission peak with the wavelength λm will be formed in the output section.

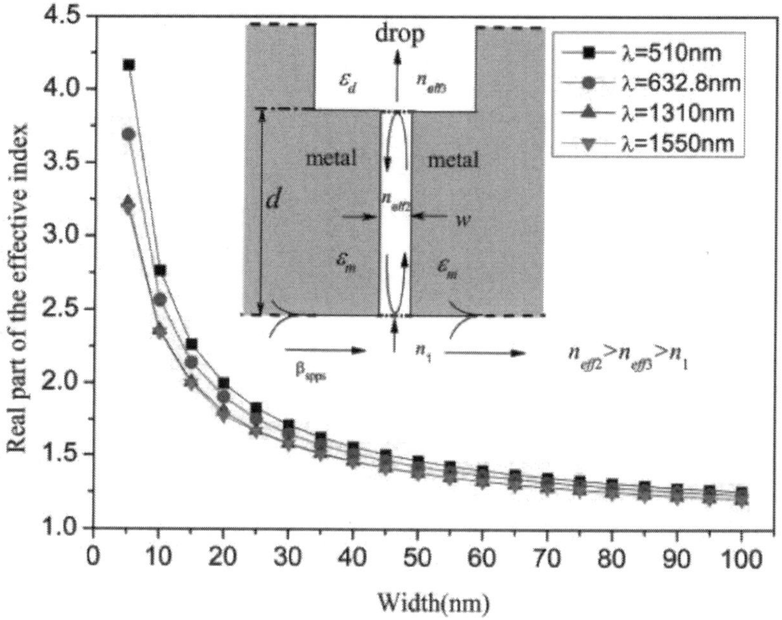

Figure 19. Dependence of real part of the effective index of SPPs in a plasmonic MDM waveguide on wavelength of the incident light and width w. Inset: schematic picture of a MDM nano-capillary resonator.

Fig. 20(a) shows a typical schematic of a 1×3 wavelength demultiplexing structure based on MDM nano-capillary resonators. The wavelength demultiplexing structure consists of three nano-capillary resonators perpendicularly connected to a bus waveguide. w_1 and d_1 stand for the width and the length of the first nano-capillary, respectively. Since the width of the bus waveguide is much smaller than the operating wavelength in the structure, only the excitation of the fundamental waveguide mode is considered. The incident light used to excite SPP wave is a TM-polarized (the magnetic field is parallel to y axis) fundamental mode. In the following FDTD simulation, the grid sizes in the x and the z directions are chosen to be Δx = 5 nm, Δz = 1.5 nm. Power monitors are respectively set at the positions of P and Q to detect the incident power of P*in* and the transmitted power of P*out*. The transmittance is defined to be T=Pout/Pin.

The width w′ of the bus waveguide is set to be 250 nm while the length of L_1 and L_2 are fixed to be 50 nm and 500 nm. As an example, three nano-capillaries have been designed to split the first, the second and third optical transmission windows, although more nano-capillaries can be added. The parameters of the structure are set to be w = 15 nm, w_1 = 250 nm, d_1 = 202 nm, d_2 = 290 nm, and d_3 = 347 nm in calculation. Fig. 20(b) shows the transmission spectra at the outputs of the three channels, and inset of fig. 20(b) shows transmittance and reflectance of the bus waveguide. From it, one can see channels l-3 can select 980 nm, 1310 nm, 1550 nm bands, respectively, and the maximum transmittance in three bands can exceed 30% (-5.2 dB). And there is also another high transmission in channel 3 around 820 nm wavelength for m = 2. Given the total phase shift φr, one can estimate the resonance wavelength from Eq. (10). Submitting λm = 1310 nm into Eq. (10) gives φr = 0.35 for d = 290 nm and $n\mathit{eff}$ = 2.01. Other resonance wavelengths can be approximately calculated with the formula. For the lengths of the nano-capillaries of 347 nm and 202 nm, resonance wavelengths are simply estimated to be 1559 nm and 926 nm. The deviation between FDTD simulation and the result from Eq. (10) could be partly attributed to the neglecting of wavelength dependence of φr. And it is partly due to the fact that Eq. (10) is derived based on the effective index approximation that SPP waves with the phase factor of $\exp(i2\pi n_{eff}x/\lambda)$ travel back and forth within a capillary, similar to a 3-dimentional plane wave with $\exp(i2\pi nx/\lambda)$ traveling in a bulk medium with refractive index n.

The FWHM of channel 1-3 are 75 nm, 130 nm, 160 nm, respectively. Obviously, the FWHM of the channel 2 and channel 3 are larger than that of channel 1. The reason is that, from the calculation inFig. 19, the effective index at short wavelength with a fixed width of nano-capillary is higher compared with the one at long wavelength, thus the waves at short wavelength have a higher reflectivity at two ends of nano-capillary and its Q factor is higher. Cross-talk is defined as the ratio between the power of the undesired and desired bands at the outputs. The cross-talk between channel 1 and channel 2 is around -19.7 dB for the 980 nm branch, and the cross-talk between them is -13.1 dB for the 1310 nm branch. The cross-talk between channel 1 and the whole channel 3 is around -19.2 dB for the 980 nm branch, and is -16.6 dB for the 1550 nm branch, although there is also another high transmission in channel 3 around 820nm wavelength for m=2. Therefore, this structure is suitable for wideband wavelengths demultiplexing.

Equation (10) indicates that the transmission behavior of each nano-capillary (channel of the demultiplexing structure) mainly depends on two parameters: the length of the nano-capillary, and the effective index of SPPs in the nano-capillary, which is determined by its width. Figure 21 shows the central wavelength of the nano-capillary resonator as a function of nano-capillary length d. One can see that the central wavelength of nano-capillary shifts toward longer wavelengths with the increasing of nano-capillary length d, as expected from equation (10). Therefore, one can realize the demultiplexing function at arbitrary wavelengths through the nano-capillary resonator by means of properly choosing the parameters of the structure, such as nano-capillary length and width.

Finally, Figure 22 shows the propagation of field H_y for two monochromatic waves with different wavelengths of 980 nm and 1550 nm launched into nano-capillary resonator demultiplexing structure. The demultiplexing effect is clearly observed. From the figure, one can see the wave with wavelength of 980 nm passing through the first nano-capillary and the wavelength of 1550 nm wave transmitting from the third nano-capillary. This is in good agreement with the transmission spectra shown in Fig. 20(b).

7. CONCLUSION

In this chapter, we present our work on nano-plasmonic waveguide filters based on tooth/teeth-shaped and nano-capillary structures. We firstly investigated a novel plasmonic waveguide filter constructed with a MDM structure engraved single rectangular tooth. The filter is of an ultra-compact size with a few hundreds of nanometers in length, with reducing fabrication difficulties, compared with previous grating-like heterostructures with a few micrometers in length. We then extended it to symmetric/asymmetric multiple-teeth, capillary structures. The asymmetrical multiple-teeth structure and the capillary structure can achieve selective narrow-band filtering and wavelength demultiplexing functions, respectively. The plasmonic filters might become a choice for the design of all-optical high-integrated architectures for optical computing and communication in nanoscale. In the future, it will be very interesting and usefully to find some solutions to improve the performance of the MIM/MDM plasmonic components. Such as to combine surface plasmons with electrically and optically pumped gain media such as semiconductor quantum dots, semiconductor quantum well, and organic dyes embedded to the dielectric part. Electrically and optically pumped semiconductor gain media and the emerging technology of graphene are also expected to provide loss compensation from visible to terahertz spectra range [26].

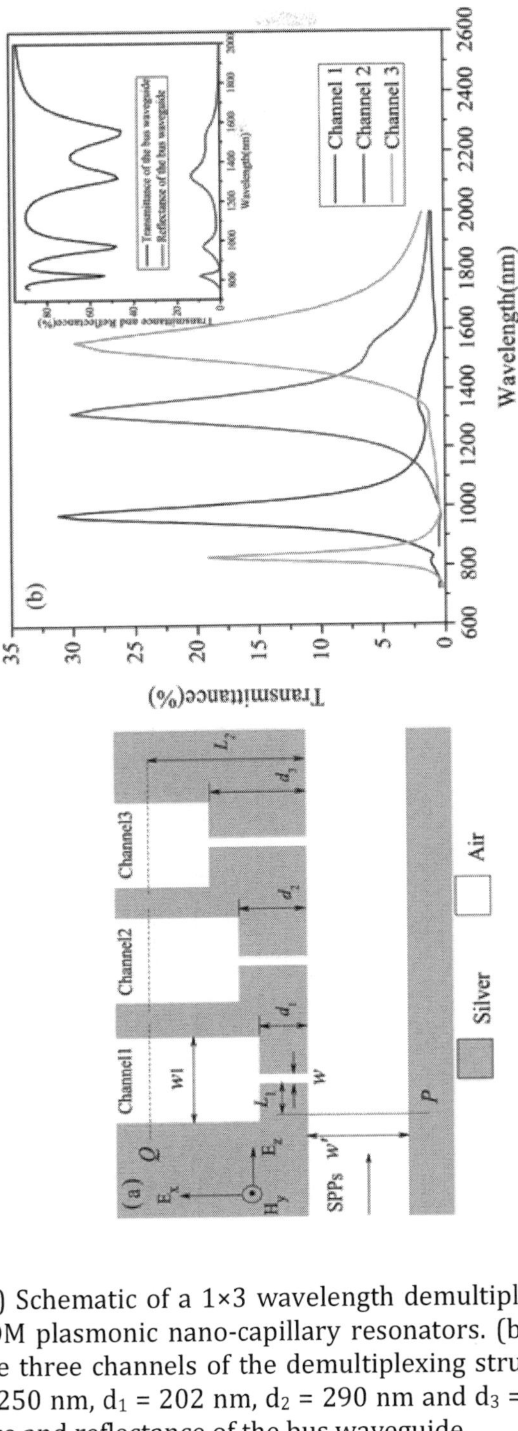

Figure 20. a) Schematic of a 1×3 wavelength demultiplexing structure based on MDM plasmonic nano-capillary resonators. (b) Transmission spectra of the three channels of the demultiplexing structure with w = 15 nm, w_1 = 250 nm, d_1 = 202 nm, d_2 = 290 nm and d_3 = 347 nm. Inset: Transmittance and reflectance of the bus waveguide.

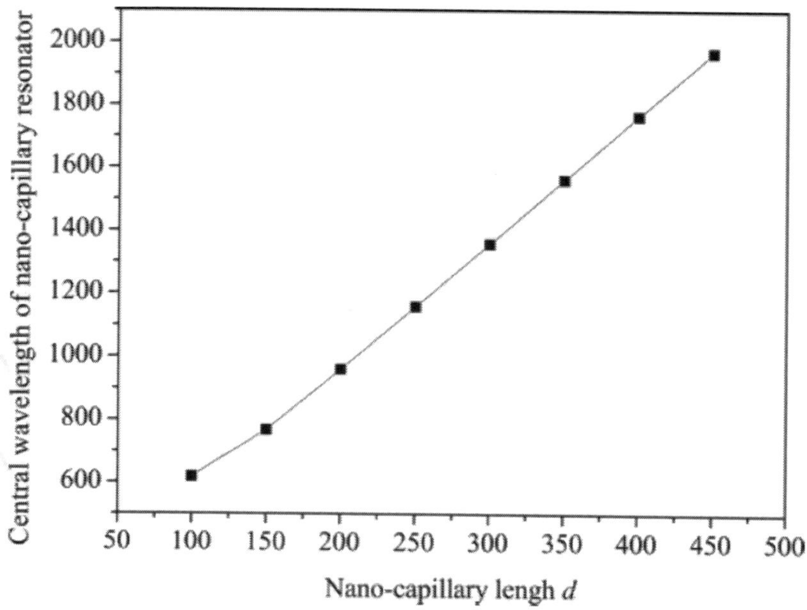

Figure 21. The central wavelength of nano-capillary resonator as a function of nano-capillary length d.

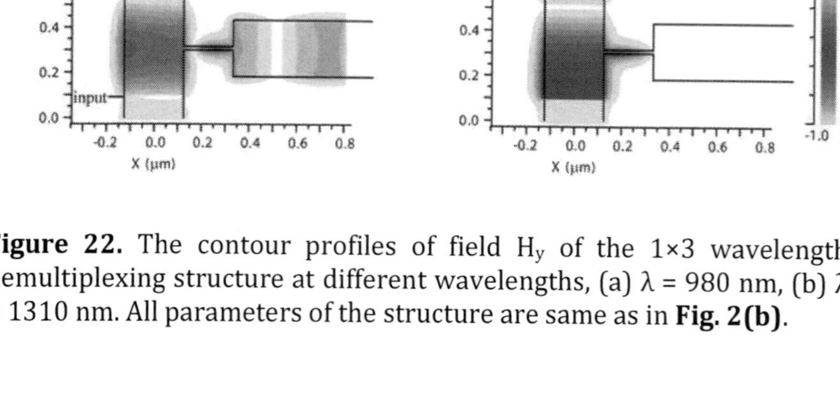

Figure 22. The contour profiles of field H_y of the 1×3 wavelength demultiplexing structure at different wavelengths, (a) λ = 980 nm, (b) λ = 1310 nm. All parameters of the structure are same as in **Fig. 2(b)**.

ACKNOWLEDGEMENT

The authors acknowledge the financial support from the National Natural Science Foundation of China (Grant No. 11977866).

REFERENCES

1. J. Davis, R. Venkatesam, A. Kaloyeros, M. Meylansky, S. Souri, K. Banerjee, K. Saraswat, A. Rahman, R. Reif, J. Meidl, ". Interconnect, on. Limits, Integration. . G. S. I. Gigascale, the. in, Century,". 21st, of. Proceedings, I. E. E. E. the, Vol, 2001

2. J. A. Conway, S. Sahni, T. Szkopek, ". Plasmonic, versus. interconnects, interconnects. a. conventional, of. comparison, crosstalk. latency, costs,". energy, Opt, Express, 152007

3. P. Bhattacharya, Optoelectronic. Semiconductor, Second. Devices, ©. Edition, 2005Prentice-Hall of India, New Dehli- 110 001. pg. 102.

4. E. Yablonovitch, ". Inhibited, Emission. Spontaneous, Solid. in-State, Physics, Phys. Electronics,", Rev. Lett 58, 201987

5. S. John, ". Strong, of. localization, in. photons, disordered. certain, superlattices,". dielectric, Phys, Rev. Lett, 581987

6. S. A. Maier, al. ". et, Local detection of electromagnetic energy transport below the diffraction limit in metal nanoparticle plasmon waveguides," Nature Mater., 22003

7. W. Saj, ". F. D. T. D. simulations, . D. of, waveguide. plasmon, silver. on, in. nanorods, lattice,". hexagonal, Opt, Express, 132005

8. S. I. Bozhevolyni, V. S. Volkov, E. Devaux, T. W. Ebbesen, ". Channel, guiding. plasmon-polariton, subwavelength. by, grooves,". metal, Phys, Rev. Lett., 952005

9. K. Tanaka, M. Tanaka, T. Sugiyama, ". Simulation, practical. of, optical. nanometric, based. circuits, surface. on, polariton. plasmon, waveguides,". gap, Opt, Express, 132005

10. J. R. Krenn, al. ". et, Non-diffraction-limited light transport by gold nanowires," Europhys Lett., 602002

11. B. Wang, G. Wang, ". Plasmon, reflectors. Bragg, on. nanocavities, metallic. flat, Appl. surface,", Phys. Lett., 872005

12. A. Boltasseva, S. I. Bozhevolnyi, T. Nikolajsen, K. Leosson, ". Compact, gratings. Bragg, Long. for-Range, plasmon. surface, J. polaritons,", Lightwave Technol. 242006

13. A. Hossieni, Y. Massoud, ". A. low-loss, plasmonic. metal-insulator-metal, reflector,". bragg, Opt, Express, 142006

14. A. Hosseini, H. Nejati, Y. Massoud, ". Modeling, methodology. design, metal-insulator-metal. for, Bragg. plasmonic, Opt. reflectors,", Express, 162008

15. J. Park, H. Kim, B. Lee, ". High, plasmonic. order, reflection. Bragg, the. in, waveguide. metal-insulator-metal, grating,". Bragg, Opt, Express, 162008

16. Z. Han, E. Forsberg, S. He, ". Surface, Bragg. plasmon, formed. gratings, metal-insulator-metal. in, I. E. E. E. waveguides,", Technol. Photon, Lett, 192007

17. Xian Shi Lin and Xu Guang Huang, "Tooth-shaped plasmonic waveguide filters with nanometeric sizes," Opt.Lett, 332008

18. J. A. Dionne, L. A. Sweatlock, H. A. Atwater, ". Plasmon, waveguides. slot, chip-scale. Towards, with. propagation, localization,". subwavelength-scale, Phys, Rev. B., 732006

19. E. D. Palik, Handbook of optical constants of solids (Academic Press, New York, NY 1985).

20. H. A. Haus, Waves and Fields in Optoelectronics (Prentice-Hall, Englewood Cliffs, NJ, 1984).

21. J. Tao, Guang. Xu, X. S. Huang, et. Lin, ". al, research. Systematical, characteristics. on, double-sided. of, nanoplasmonic. teeth-shaped, filters,". J. waveguide, Opt. Soc. Am. B, 272010

22. X. S. Lin, X. G. Huang, ". Numerical, of. a. modeling, nanoplasmonic. teeth-shaped, filter,". J. waveguide, Opt. Soc. Am. B, 262009

23. J. Tao, X. G. Huang, X. S. Lin, Q. Zhang, X. P. Jin, ". A. narrow-band, plasmonic. subwavelength, filter. waveguide, asymmetrical. with, structure,". multiple-teeth-shaped, Opt, Express, 172009

24. Qin Zhang, Xu-Guang Huang, Xian-Shi Lin, Jin Tao, and Xiao-Ping Jin, "A subwavelength coupler-type MIM optical filter," Opt. Express, Vol.17200975497555

25. Jin Tao, Xu Guang Huang, and Jia Hu Zhu, "A wavelength demultiplexing structure based on metal-dielectric-metal plasmonic nano-capillary resonators," Opt.Express, 182010

26. A. A. Dubinov, V. Y. Aleshkin, V. Mitin, T. Otsuji, V. Ryzhii, ". Terahertz, plasmons. surface, optically. in, graphene. pumped, J. structures,", Phys, Phys.Condens. Matter, 232011

7

ANALYSIS OF EFFICIENTLY POLED ELECTRO-OPTIC POLYMER/TIO$_2$ VERTICAL SLOT WAVEGUIDE MODULATORS

Y. Enami[a], H. Nakamura[a], J. Luo[b], A.K-Y. Jen[b]

a School of System Engineering, Kochi University of Technology, Kochi 782-8502, Japan

b Department of Materials Science and Engineering, University of Washington, Seattle, WA 98198, USA

ABSTRACT

We analyze the advantages of an electro-optic (EO) polymer/TiO_2 vertical slot waveguide modulator based on a low-index EO polymer (SEO125). This modulator can realize a lower half-wave voltage ($V\pi$)-electrode length (L_e) product ($V\pi Le$) when compared with hybrid EO polymer (EOP)/sol–gel silica waveguide modulators because of the high mode confinement of the guided light and the high poling efficiency. We show the enhancement of the poling efficiency in these devices when the EO polymers are poled with TiO_2 and sol–gel silica layers. We also enhance the EO coefficient to a level of 260 pm/V at a wavelength of 1.31 μm for a high-index EOP (SEO100) deposited on TiO_2, a sol–gel silica cladding layer, and an additional interfacial layer.

KEYWORDS

Electro-optic modulators; Electro-optic polymers; Vertical TiO_2 slot waveguide; Sol–gel silica cladding

1. INTRODUCTION

Polymer modulators have been the only optical modulators to have bandwidths of >60 GHz, with a widest reported bandwidth of 113 GHz, because of the low dielectric dispersion of the electro-optic polymers (EOPs) used in these devices [1]. We have previously demonstrated a low half-wave voltage (*Vπ*) in the 0.65–1.0 V range [2] and [3]with relatively low optical propagation loss of 5 dB/cm, using a hybrid EOP/sol–gel (SG) silica waveguide modulator with an in-device electro-optic (EO) coefficient of 142 pm/V at a wavelength of 1550 nm. The *VπLe* product (where *Vπ* is the half-wave voltage, and *Le* is the electrode length) is considered to be an appropriate figure of merit (FOM) for a complementary metal-oxide-semiconductor (CMOS)-compatible integrated optical modulator for optical interconnections based on Si modulators [4], [5], [6], [7] and [8]. Only a few reports on Si modulators have focused on the optical losses for realistic applications. A combination of the FOM and the optical loss can be defined as a more realistic FOM as the *VπLoss* product [V dB], in which *VπLe* [V cm] was multiplied by the optical propagation loss [dB/cm] in the active region (phase shifter). Our hybrid polymer modulator has a *VπLe* product of 1.56 V cm and a *VπLoss* product of 7.8 V dB in the hybrid EOP/SG silica waveguide modulator [2]. To the best of the authors' knowledge, one of the best reported Si modulators to date showed a *VπLe* product of 0.78 V cm and a*VπLoss* product of 6.7 V dB [4], although other Si modulators showed *VπLe* products of 1–3 V cm with higher *VπLoss* products of >27 V dB [5], [6], [7] and [8]. Other CMOS compatible modulators include the III–V electro-absorption modulator (EAM) with a bandwidth of 52 GHz [9], EOP/Si slot waveguides, and EOP/Si photonic crystal (PC) waveguide modulators [10], [11], [12], [13] and [14], and the properties of these devices are summarized in Table 1.

EOP/Si slot waveguide modulators were demonstrated based on coplanar Si slot waveguides [11], [12] and [13], in which no optical modes exist in the high-index Si; instead, the waveguide mode was confined within the low index EOP between the Si slots, because the Si size is critically reduced to a less than single mode condition (*λ/2neffective*), where *λ* is the wavelength and *neffective* is the effective refractive index of the waveguide. EOP/Au slot waveguide modulators were demonstrated using Au plasmonic slot waveguides. For actual application of these devices to optical interconnections, the optical propagation loss must be further reduced, and another FOM, the *VπLoss* product, must also be considered. We have demonstrated an all-dielectric EOP/TiO_2 multilayer vertical slot waveguide modulator for the first time [15] that provides higher poling efficiency, a wider EO modulation bandwidth and lower optical propagation losses without using highly-doped Si layers that limit device performance. We reduced the $V\pi Le$product to 2 V cm and the $V\pi Loss$ product to 14 V dB by using low-index EOP (SEO125, with refractive index of 1.621, and in-device EO coefficient r_{33}=78 pm/V at a wavelength of 1.55 μm), based on the enhanced conductivity of the SG silica under a cladding layer[16]. In this modulator, the enhanced conductivity of the SG silica under cladding realized efficient poling of the EOP with a selective TiO_2 layer. We also demonstrated high EO coefficients of 226 and 198 pm/V at wavelengths of 1.31 and 1.55 μm, respectively, for an efficiently poled high refractive index EOP film of SEO100 deposited on a TiO_2 selective layer and SG silica cladding [17]. In this paper, we analyze the advantages of EOP/TiO_2slot waveguide modulators for further reduction of the $V\pi Le$ product when compared with hybrid EOP/SG silica waveguide modulators.

Table 1. CMOS-compatible optical modulators.

Type	$V_\pi L$ [V cm]	Loss	$V_\pi Loss$ [V dB]	3 dB-BW [GHz]	r_{33}	Ref.
EOP/SG	1.56	5	7.8		142	2
Si	0.78	6.6	6.7		–	4
Si	1	31	31		–	5
Si	1.4	19	27	12(1 mm) 30 (0.25 mm)	–	6
Si	3.5	12	42	15	–	7
Si	2.7	10	28	26	–	8
III–V (EAM)	0.022	49	52	74	–	9
EOP/Si PC	0.029	–	–		89	10
EOP/Si slot	0.5	–	–		30	11
EOP/Si slot	0.052	60	3.12	10	230	12
EOP/Si slot	1.1	60	66		18	13
EOP/Au slot	–	4000	–		21	14
EOP/TiO_2 slot	2	7	14		78	16

Loss: Optical loss [dB/cm].

2. EO POLYMER/TIO_2 SLOT WAVEGUIDE MODULATORS

2.1. Analysis and demonstration of EO polymer/TiO_2 vertical slot waveguide modulators

Mach–Zehnder-type EOP/TiO_2 multilayer slot waveguide modulators were fabricated using a 0.3–0.6-µm-thick EOP layer sandwiched between two 0.1-µm-thick TiO_2 thin film layers. A cross-sectional view of a 4-µm-wide Mach–Zehnder-type waveguide used in the modulator is shown in Fig. 1(a). A standard hybrid EOP/SG silica waveguide modulator is also shown in Fig. 1(b) [2]. The half-wave length $V\pi$ is expressed as

$$V_{\pi} = \frac{\lambda d_{effective}}{2n^3 r_{33} L_e \Gamma} \tag{1}$$

$$d_{effective} = d_{EO} + \sqrt{\varepsilon_{EO}/\varepsilon_{TiO_2}}\, d_{TiO_2} + \sqrt{\varepsilon_{EO}/\varepsilon_{CYTOP}}\, d_{CYTOP} + \sqrt{\varepsilon_{EO}/\varepsilon_{SG}}\, d_{SG} \tag{2}$$

where *deffective* is the effective electrode distance, n is the refractive index of the active material, and Γ is the mode overlap integral between the optical wave in the active region and the applied electric field. εEO (~3), ε_{TiO2} (>30), $\varepsilon CYTOP$ (~2), and εSG (~3) are the dielectric constants for the EOP, TiO_2, CYTOP and SG layers, respectively.dEO, d_{TiO2}, and $dCYTOP$ are the thicknesses of the EOP, TiO_2 and CYTOP layers, respectively. Because the TiO_2 layer thickness is 0.1 µm, the term d_{TiO2} can be neglected. The thicknesses of the SG silica layer under cladding and the CYTOP layer were 4 µm and 1.2 µm, respectively. The refractive indices of the SEO125 EOP, TiO_2, CYTOP, and SG silica cladding layers were 1.621, 2.567, 1.328, and 1.487, respectively. As a first step, the mode profile in the slot waveguide was calculated using the three-dimensional finite-difference time domain (3D FDTD) method. One mode profile for the thickness of the EOP (tEO) of 0.5 µm was shown in Fig. 2. $\Gamma/d_{eff}$with respect to the tEO was obtained from the result for the profile as shown in Fig. 3.

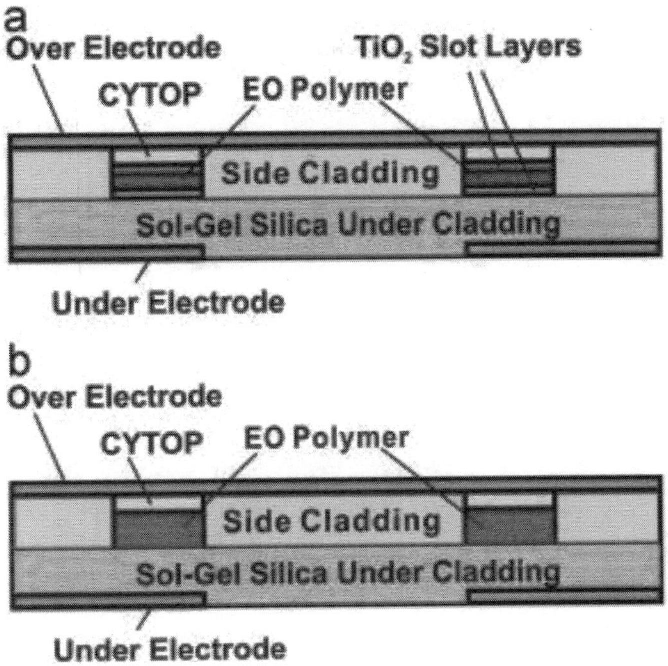

Figure. 1. Schematic cross-sectional views of active regions. (a) EOP/TiO_2 multilayer vertical slot waveguide modulator. (b) Hybrid EOP/sol–gel silica waveguide modulator.

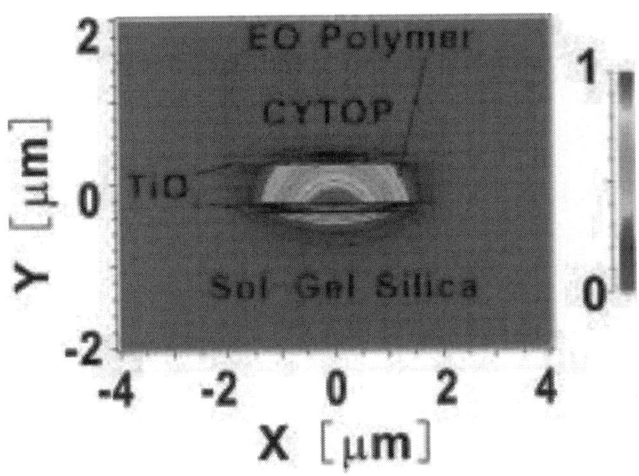

Figure. 2. One mode profile for the thickness of the EOP (*tEO*) of 0.5 µm in the electro-optic polymer/TiO_2 vertical slot waveguide modulator.

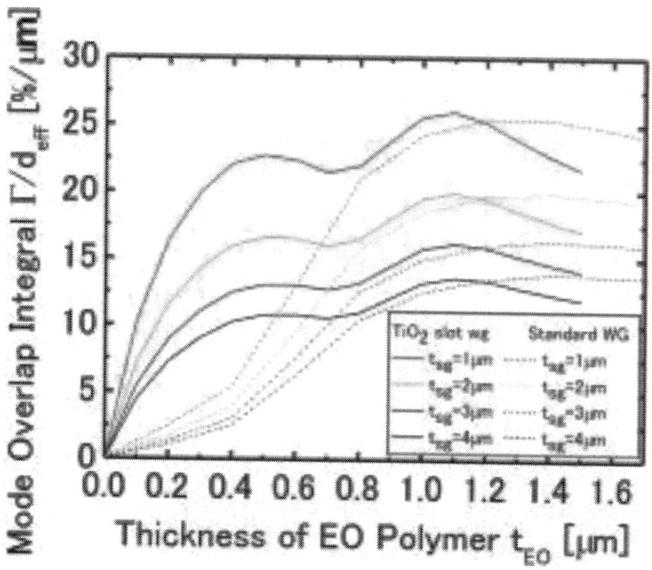

Figure. 3. Normalized mode overlap integral Γ between the optical mode and the applied electric field divided by the effective electrode distance *deffective* with respect to the thickness of the EOP layer, *tEO*. The solid lines (EOP/TiO_2 vertical slot waveguide modulators) in red, green, blue, and black show the results for SG silica thicknesses under cladding of 1, 2, 3, and 4 μm, respectively. The dotted lines (standard hybrid EOP/SG silica waveguide modulators) in red, green, blue, and black show the results for SG silica thicknesses under cladding of 1, 2, 3, and 4 μm, respectively. (For interpretation of the references to color in this figure legend, the reader is referred to the web version of this article.)

Next, the $V\pi Le$ products of the EOP/TiO_2 slot waveguide modulators and those of standard hybrid modulators were calculated from the in-device r_{33} value of 70 pm/V and the obtained value of Γ. The dependence of the $V\pi Le$ product for the slot waveguide modulator (see Fig. 1(a)) on the EOP thickness was obtained for 1-μm-thick and 4-μm-thick SG silica under cladding layers, as shown in Fig. 4. The same dependence of the$V\pi Le$ product for a standard hybrid modulator (see Fig. 1(b)) was also obtained for comparison. When *tsg* is 4 μm, the slot waveguide modulator does not show any significant advantage for >0.75-μm-thick EOP layers. In this calculation, we assumed that the in-device EO coefficient r_{33} (70 pm/V) was same for both the TiO_2 slot waveguide modulators and the standard modulator. In the actual modulator devices, the in-device EO coefficient for the TiO_2 slot waveguide modulator is higher than that for the standard modulator, which resulted in a lower $V\pi Le$ product because the poling efficiency of the EOP layer was improved by the TiO_2 and SG layers [17]. It was also difficult to pole 0.3–0.6-μm-thick EO polymers without the TiO_2 and SG layers because of dielectric breakdown at lower poling voltages of <100 V at a glass transition temperature of 150 °C. When the EO polymer was deposited on the TiO_2 and SG layers, the EOP was efficiently poled without the occurrence of the breakdown, which was a major advantage of the EOP/TiO_2 slot waveguide modulators. When *tsg* is critically reduced to 1 μm, the $V\pi Le$product calculated for the EOP/TiO_2 slot waveguide modulator is two times lower than that of the modulator (*tsg* of 4 μm), as shown in Fig. 4.

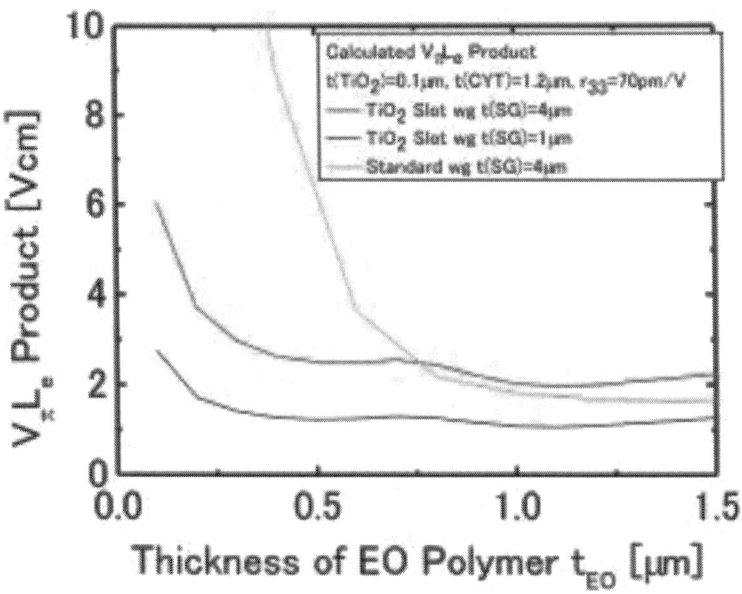

Figure. 4. $V_\pi L$ product with respect to the EOP (SEO125) thickness, as calculated by the 3D FDTD method. The thicknesses of the TiO_2 and CYTOP layers are 0.1 and 1.2 μm, respectively. r_{33}=70 pm/V. Red and blue lines show the product values for EOP/TiO_2 slot waveguide modulators with sol–gel silica layer thicknesses under cladding of 4 and 1 μm, respectively. The green line shows the product for the standard hybrid EOP/sol–gel silica waveguide modulator. (For interpretation of the references to color in this figure legend, the reader is referred to the web version of this article.)

A SG solution was prepared for the waveguide side cladding and under-cladding layers, consisting of methacryloyloxy propyltrimethoxysilane (MAPTMS) and an index modifier (zirconium (IV)-*n*-propoxide) with a molar ratio of 95(MAPTMS)/5 mol%. A 0.1-N HCl solution was used as a catalyst, and Irgacure 184 (Ciba) was used as a photoinitiator for the side cladding solutions to accelerate the hydrolysis of the silica for subsequent wet etching in isopropanol. A 4-µm-thick under-cladding layer was coated on a 100-nm-thick Au/10-nm-thick Ti lower electrode/silica (6 µm)-on-silicon substrate structure. The under-electrode was separated for dual-drive (push-pull) EO modulation. A 4-µm-thick sol–gel silica side-cladding layer was then spin-coated on the under-cladding. Mercury i-line (365 nm) radiation was delivered from a mask aligner to the sol–gel layer through a photomask to etch a 4-µm-wide core region. The regions that were exposed to radiation were then cross-linked to form a silica network. The ultraviolet-irradiated areas became insoluble in isopropanol, which was used as a wet etchant. The SG silica waveguide was hard baked at 150 °C for 1–2 h, and a 100–nm-thick TiO_2 layer was sputtered at radio frequency to form a lower slot core layer on the under-cladding between the etched SG side-cladding layers. A 0.3–0.6-µm-thick low-index EO polymer (SEO125) with 35 wt% chromophore doping in amorphous polycarbonate (APC) was then spin-coated under the TiO_2 slot layer and baked overnight at 80 °C in a vacuum oven. After the EO polymer was poled between an Au poling electrode deposited on the EO polymer and the under-electrode, the poling electrode was removed using an iodine and potassium iodine solution in deionized water. Finally, another 100-nm-thick TiO_2 slot layer was sputtered on the EO polymer, and a 1.2-µm-thick low-index perfluorinated polymer Cytop® layer was spin-coated to act as a buffer layer for consecutive deposition of an Au over-electrode.

The $V\pi Le$ product was experimentally measured at 1 kHz for several modulators with different EOP thickness (tEO) values, ranging from 0.3 to 0.6 µm, as shown in Fig. 5. The measured $V\pi Le$ products were well matched with the theoretically calculated values. When the baking time for the SG silica was reduced to 0.75 h, the $V\pi Le$ product showed its lowest value of 2.0 V cm, and the highest in-device $r33$ value of 78 pm/V was obtained for the 0.6-µm-thick SEO125 EOP at the wavelength of 1.55 µm because of the increased conductivity of the SG silica and the reduced conductance of the EOP [11]. The $V\pi Loss$ product was 14 V dB (propagation loss of 7 dB/cm). We measured a coupling loss of 7 dB/facet between a standard single mode fiber and the modulator without a mode convertor or a grating coupler when light through the fiber was butt-coupled into the modulator. Total optical insertion loss was 17 dB for the 14-mm-long MZ modulator, including a propagation loss of 10 dB (7 dB/cm) and a butt -coupling loss of 7 dB/facet between a standard SM fiber and the modulator when the fiber was butt-coupled into the slot waveguide modulator and an output light was focused on a power meter using a microscope lens. However, the possibility of breakdown was increased when the baking time was critically reduced. When the SG silica was baked for 2 h, the conductivity for the multiple TiO_2 slot and SG layers reduced. We obtained a similar in-device *r33* of 70 pm/V for EOP thicknesses in the range from 0.3 to 0.5 μm, as shown inFig. 5. When the under-cladding thickness in the hybrid modulator was reduced to <4 μm, the optical propagation loss increased to >20 dB/cm. In contrast, the EOP/TiO_2 vertical slot waveguide modulators all showed similar propagation losses (e.g., 15 dB/cm), even when *tsg* was reduced to 1.8 μm, because the mode tail in the slot waveguide was reduced. For the modulators that have 1.8-μm-thick sol–gel silica undercladding, the$V\pi L_e$ was more than 7 V cm because the poling efficiency for the EO polymer was not optimized. The propagation loss was varied and depended on the waveguide quality of the modulator samples. Even though some modulator samples did not have optimized poling efficiency, we obtained the lowest propagation loss of 3.5 dB/cm for the modulators that have 4-μm-thick sol–gel under cladding. The propagation loss was varied from 3.5 to 8.6 dB/cm for each waveguide in the same samples. The propagation loss was much lower than other Si slot waveguide modulators (e.g. 60 dB/cm) as shown in Table 1.

2.2. Enhancement of the in-device electro-optic coefficient

We also examined the $V\pi Le$ product for an EOP/single TiO_2 slot waveguide modulator, in which the EOP was poled without the TiO_2 and CYTOP buffer layers, as shown in Fig. 6(a). The poling setup is same as that used for the EOP/TiO_2 vertical slot waveguide modulators, except for the removal of the TiO_2 underlayer. Therefore, we can find the difference in poling efficiency between these structures (i.e., when poled with and without the TiO_2 layer) from the measured $V\pi$. After poling of the 0.4 μm-thick SEO125 EOP, the poling electrode was removed by a wet-etching method. Then RF-sputtered TiO_2 and CYTOP buffer layers were deposited on the EOP, and a final Au over-electrode was deposited, as shown in Fig. 6(b). This process is common for EOP/TiO_2 vertical slot waveguide modulators. The $V\pi Le$ product was 5.2 V cm ($V\pi$=10.3 V for Le=0.5 cm), and the $V\pi Loss$ product was 41.6 V dB (propagation loss of 8 dB/cm). The in-device r_{33} was 39 pm/V at a wavelength of 1.55 μm. Because the in-device r_{33} was 70–78 pm/V when the SEO125 EOP was poled on the TiO_2 underlayer and SG silica layer in the EOP/TiO_2vertical slot waveguide modulator, as shown in Fig. 1(a), the in-device r_{33} was therefore increased by a factor of 2.

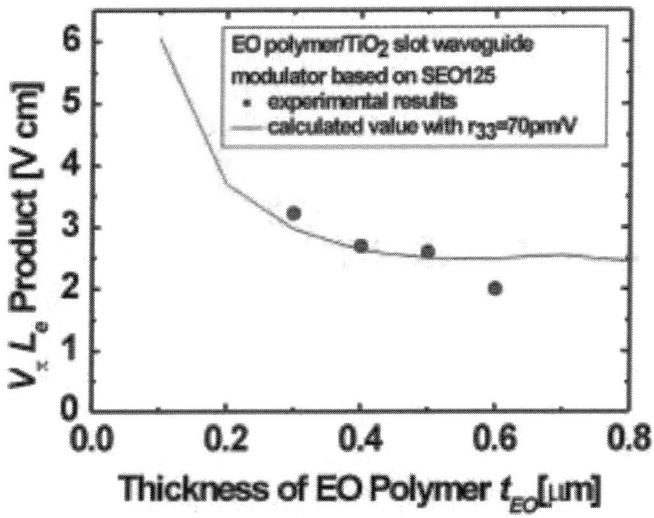

Figure. 5. $V_\pi L_e$ product with respect to the thickness of the EOP (SEO125). Blue closed circles denote the experimental results and the red line shows the result calculated by the 3D FDTD method. (For interpretation of the references to color in this figure legend, the reader is referred to the web version of this article.)

We examined the increased poling efficiency of the high-index SEO100 EOP when coated on TiO_2 and SG silica layers using an ITO glass substrate. The poling setup is the same as that used for the EOP/TiO_2 slot waveguide modulator. Because the refractive index of SEO100 was 1.702, the refractive index of the slot layer must be increased to ensure sufficient mode confinement in the EOP/TiO_2 slot waveguide modulator and to reduce the optical losses. To date, the highest reported EO coefficients of 260 and 215 pm/V at wavelengths of 1.31 and 1.55 μm, respectively, were obtained using Teng and Man's ellipsometric method [18]. When SEO100 was poled without the TiO_2 and sol–gel silica layers, dielectric breakdown occurred frequently when the poling voltage was <10 V at the glass transition temperature of 150 °C. The EO coefficient for poled SEO100 on TiO_2 and SG silica layers was three times higher than that for SEO125. When we successfully use SEO100 in the modulator, the $V\pi Le$ product would be reduced by a factor of six with the reduced thickness of the SG silica and the enhanced EO coefficient, which will then result in a $V\pi Le$ product of less than 0.3 V cm and a $V\pi Loss$ product of 2.3 V dB.

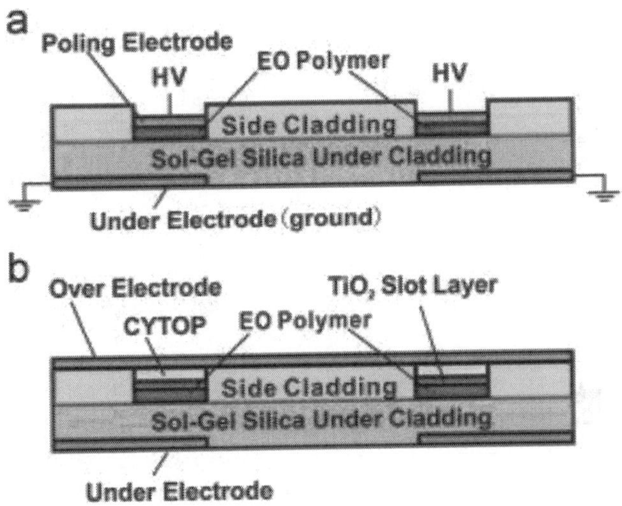

Figure. 6. Schematic cross-sectional view of the active region of the EOP/single TiO_2 slot waveguide modulator. (a) Poling setup for the modulator without TiO_2 underlayer. (b) The modulator fabricated after the poling.

3. CONCLUSION

We have analyzed EOP/TiO_2 vertical slot waveguide modulators and have examined the poling efficiency in these modulators. The in-device EO coefficient was successfully enhanced in the devices when the EOP was poled with TiO_2 and SG layers.

ACKNOWLEDGMENT

This work was supported by a Grant-in-Aid for Scientific Research (A) (Grant nos.21246060 and 24246063) from MEXT, Japan, the Air Force

Office of Scientific Research, United States (FA8650-12-M-5130 under the Small Business Technology Transfer Research (STTR) program), and Soluxra, LLC, USA.

REFERENCES

1. D. Chen, H.R. Fetterman, A. Chen, W.H. Steier, L.R. Dalton, W. Wang, Y. Shi **Demonstration of 110 GHz electro-optic polymer modulators** Appl. Phys. Lett., 70 (1997), pp. 3335–3337

2. Y. Enami, D. Mathine, C.T. DeRose, R.T. Norwood, J. Luo, A.K.-Y. Jen, N. Peyghambarian **Hybrid cross-linkable polymer/sol–gel waveguide modulators with 0.65 V** Appl. Phys. Let, 91 (2007), p. 093505

3. Y. Enami, C.T. DeRose, D. Mathine, C. Loychik, C. Greenlee, R.A. Norwood, T.D. Kim, J. Luo, Y. Tian, A.K.-Y. Jen, N. Peyghambarian **Hybrid polymer/sol–gel waveguide modulators with exceptionally large electro-optic coefficients** Nat. Photonics, 1 (2007), pp. 180–185

4. H. Goykhman, B. Desiatov, S. Ben-Ezra, J. Shappir, U. Levy **Optimization of efficiency-loss figure of merit in carrier-depletion silicon Mach–Zehnder optical modulator** Opt. Express, 21 (2013), pp. 19518–19529

5. M.R. Watts, W.A. Zortman, D.C. Trotter, R.W. Young, A.L. Lentine **Low-voltage, compact, depletion-mode, silicon Mach–Zehnder modulator** IEEE J. Sel. Top. Quantum Electron., 16 (2010), pp. 159–164

6. N.-N. Feng, S. Liao, D. Feng, P. Dong, D. Zheng, H. Liang, R. Shafiha, G. Li, J.E. Cunningham, A.V. Krishanamoorthy, M. Asghari **High speed carrier-depletion modulators with 1.4 V-cm $V_\pi L$ integrated on 0.25 μm silicon-on-insulator waveguides** Opt. Express, 18 (2010), pp. 7994–7999

7. M. Ziebell, D. Marris-Morini, G. Rasigade, J.-M. Fédéli, P. Crozat, E. Cassan, D. Bouville, L. Vivien **40 Gbit/s low-loss silicon optical modulator based on a pipin diode** Opt. Express, 20 (2012), pp. 10591–10596

8. X. Tu, T.-Y. Liow, J. Song, X. Luo, Q. Fang, M. Yu, G.-Q. Lo **50-Gb/s silicon optical modulator with traveling-wave electrodes** Opt. Express, 21 (2013), pp. 12776–12782

9. Y. Tang, J.D. Peters, J.E. Bowers **Over 67 GHz bandwidth hybrid silicon electroabsorption modulator with asymmetric segmented electrode for 1.3 μm transmission** Opt. Express, 20 (2012), pp. 11529–11535

10. X. Zhang, A. Hosseini, S. Chakravarty, J. Luo, A.K.-Y. Jen, R.T. Chen **Wide optical spectrum range, subvolt, compact modulator based on an electro-optic polymer refilled silicon slot photonic crystal waveguide** Opt. Lett., 38 (2013), pp. 4931–4934

11. T. Baehr-Jones, B. Penkov, J. Huang, P. Sullivan, J. Davies, J. Takayesu, J. Luo, T.-D. Kim, L. Dalton, A. Jen, M. Hochberg, A. Scherer **Nonlinear polymer-clad silicon slot waveguide modulator with a half wave voltage of 0.25 V** Appl. Phys. Lett., 92 (2008), p. 163303

12. R. Palmer, S. Koeber, D.L. Elder, M. Woessner, W. Hemi, D. Korn, M. Lauermann, W. Bogaerts, L. Dalton, J. Leuthold **High-speed, low drive-voltage silicon-organic hybrid modulator based on a binary-chromophore electro-optic material** J. Lightwave Tech., 16 (2014), pp. 2726–2734

13. L. Alloatti, R. Palmer, S. Diebold, K.P. Pahl, B. Chen, R. Dinu, M. Fournier, J.-M. Fedeli, T. Zwick, W. Freude, C. Koos, J. Leuthold **100 GHz silicon–organic hybrid modulator** Light: Sci. Appl., 3 (2014), p. e173

14. Melikyan, L. Alloatti, A. Muslija, D. Hillerkuss, P.C. Schindler, J. Li, R. Palmer, D. Korn, S. Muehlbrandt, D. Van Thourhout, B. Chen, R. Dinu, M. Sommer, C. Koos, M. Kohl, W. Freude, J. Leuthold **High-speed plasmonic phase modulators** Nat. Photonics, 8 (2014), pp. 229–233

15. Y. Enami, B. Yuan, M. Tanaka, J. Luo, A.K.-Y. Jen **Electro-optic polymer/TiO_2 multilayer slot waveguide modulators** Appl. Phys. Lett., 101 (2012), p. 123509

16. Y. Enami, Y. Jouane, J. Luo, A.K.-Y. Jen **Enhanced conductivity of sol–gel silica cladding for efficient poling in electro-optic**

polymer/TiO_2 vertical slot waveguide modulators Opt. Express, 22 (2014), pp. 30191–30199

17. Y. Jouane, Y.-C. Chang, D. Zhang, J. Luo, A.K.-Y. Jen, Y. Enami **Unprecedented highest electro-optic coefficient of 226 pm/V for electro-optic polymer/TiO_2multilayer slot waveguide modulators** Opt. Express, 22 (2014), pp. 27725–27732

18. C.C. Teng, H.T. Man **Simple reflection technique for measuring the electro-optic coefficient of poled polymers** Appl. Phys. Lett., 56 (1990), p. 1734

8

FLUIDIC OPTICAL DEVICES BASED ON THERMAL LENS EFFECT

Hong Duc Doan and Kazuyoshi Fushinobu

[1] Department of Mechanical and Control Engineering, Tokyo Institute of Technology, Meguro-ku, Tokyo,, Japan

1. INTRODUCTION

Gordon et al. [1] reported that the beam shape of incident laser light expands after passing through a liquid medium. This phenomenon was termed "the thermal lens effect," and it has become a well-known photo-thermal phenomenon. Phenomenological, optical, and spectroscopic studies of the thermal lens effect have been carried out to describe nonlinear defocusing effect [2-8]. Recent progress in laser technology has revealed the various aspects of the thermal lens effect. Based on these efforts, other mechanisms, such as liquid density, electronic population, and molecular orientation, have been found to play important role as well as thermal lens effect. Recent studies term these effects as "the transient lens effect" [9,10]. The main advantage of using the transient lens effect in Photo-Thermal-Spectroscopy is that the sensitivity is 100 to 1000 greater than a traditional absorptiometry [11].

In this research, a new idea of applying the thermal lens effect in order to develop fluidic optical device is proposed. A schematic of the concept is shown in Fig. 1. A rectangular solid region shown inFig. 1a represents the liquid medium, which has a temperature field generated by a heater-heat sink system or laser-induced absorption. By controlling the temperature field as well as the refractive index distribution of the liquid medium, the refractive angle of each light ray passing through the liquid medium can be controlled in order to develop fluidic optical devices such as: an optical switching inFig. 1a to change the direction of the input laser beam, a laser beam shaper in Fig. 1b to transform a Gaussian beam to a flat-top beam and a fluidic divergent lens in Fig. 1c. Merits of these devices include flexibility of optical parameters, versatility and low cost.

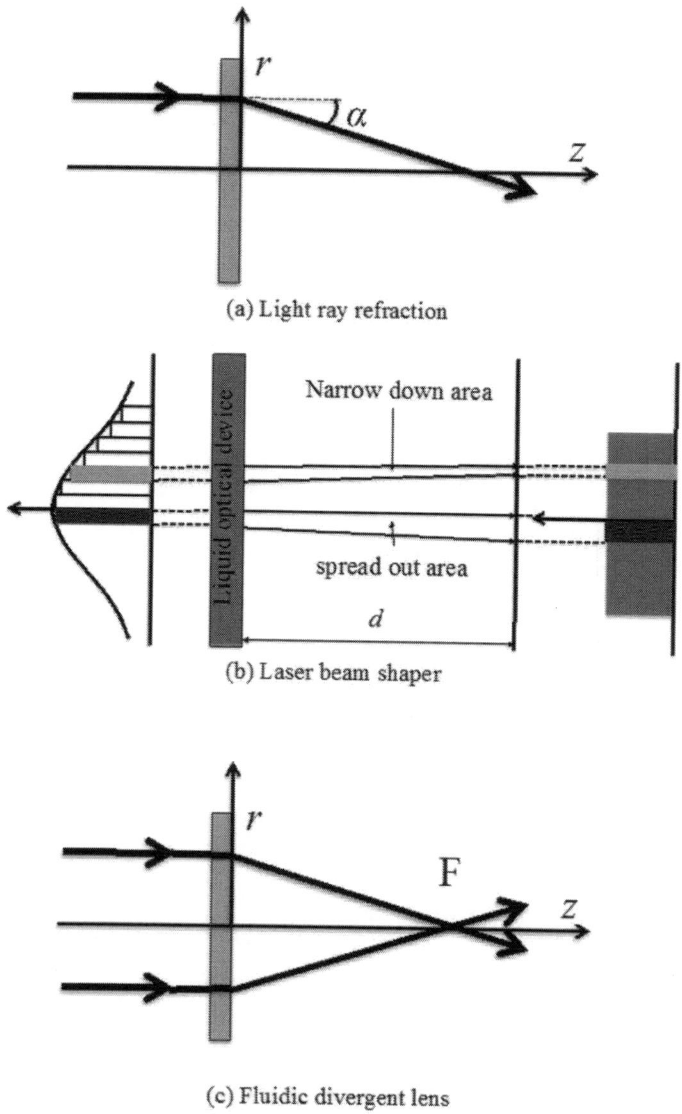

(a) Light ray refraction

(b) Laser beam shaper

(c) Fluidic divergent lens

Figure 1. Schematic of the concept of the fluidic optical devices

2. FUNDAMENTAL LIGHT RAY TRANSMITTED IN ONE-DIMENSIONAL REFRACTIVE INDEX MEDIUM

In this section, as a first step to develop fluidic optical device, the refractive characteristics of a probe beam, which is transmitted in one-dimensional temperature distribution in a liquid medium is presented.

2.1. Theoretical Background

The light ray is modeled in the domain shown in Fig. 2 in order to calculate the refractive angle of the probe beam, which is transmitted in a one-dimensional temperature distribution in the liquid medium. The light ray direction transmitted in a medium having a refractive index dependent only on the y-axis, is described by the following form [12]

$$x = \int_0^y \frac{n_0 \sin\theta \cos\varphi}{\sqrt{n^2(y) - n_0^2 \sin^2\theta}} dy \tag{1}$$

$$z = \int_0^y \frac{n_0 \sin\theta \sin\varphi}{\sqrt{n^2(y) - n_0^2 \sin^2\theta}} dy \tag{2}$$

In which, the light ray passes through the medium at coordinate center, θ and φ are the incident angle with y and x-axis respectively, n_0 is the refractive index of the medium at the coordinate center.

Figure 2 shows a schematic diagram of the model set up. The rectangular solid medium in the figure represents the domain considered in the calculation which consists of ethylene glycol. In the medium, ethylene glycol has linear temperature distribution in only y-axis direction and the probe beam propagate along z-axis direction. Therefore, $\theta = \varphi = \pi/2$, and Eq. (1), (2) become:

$$x = 0 \tag{3}$$

$$z = \int_0^y \frac{n_0}{\sqrt{n^2(y) - n_0^2}} dy \tag{4}$$

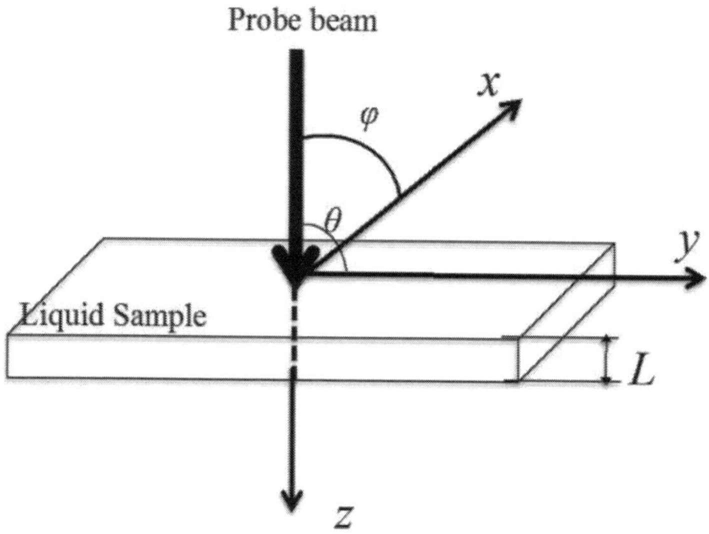

Figure 2. Schematic diagram of computational domain

The temperature distribution of the liquid medium is modeled with a linear function of the variable *y*and temperature *T* at point *y* is calculated following:

$$T(y) = T_0 + \frac{dT}{dy} y \tag{5}$$

In which, T_0 is the temperature of the liquid medium at coordinate origin and dT/dy is constant.

Furthermore, between 0°C and 100°C refractive index of ethylene glycol is a linear function of temperature with refractive index change dn/dT = -2.6×10^{-4} 1/K [13]. Therefore, the relationship between refractive index and the variable *y* can be rewriten as follows:

$$y = \frac{n(y) - n_0}{dn / dy} \tag{6}$$

and

$$dy = dn \times \frac{dy}{dn} = dn \times \frac{dy}{dT} \times \frac{dT}{dn} = \frac{dn}{k} \tag{7}$$

Where,

$$k = -2.6 \times 10^{-4} \times \frac{dT}{dy} \tag{8}$$

By substituting Eq. (7) into Eq. (4) and solving the differential equation, we can obtain the relationship between y and z as:

$$y = \frac{n_0[\exp(kz/n_0) - 1]^2}{2k\exp(kz/n_0)} \tag{9}$$

And refractive angle (RA) can be obtained as:

$$\alpha = \frac{dy}{dz} = \frac{1}{2}\exp(\frac{kz}{n_0}) - \frac{1}{2}\exp(-\frac{kz}{n_0}) \tag{10}$$

Equation (10) shows the expression of the refractive angle as a function of the temperature gradient and the thickness of the liquid medium (optical path length). Figure 3 shows the relationship between refractive angle and temperature gradient at points where the thickness of sample, L, is 1.5, 3.0 and 4.5 mm respectively. As shown in Fig. 3 the relationship between the refractive angle and temperature gradient can be well approximated as linear at a small temperature gradient.

2.2. Experimental Set-Up

Figure 4 shows the experimental set-up to measure the refractive angle. Fluidic optical device in Fig. 4is a pyrex vessel (internal size: $21\times10\times L$ [mm], thickness of the liquid medium, L, can be varied) filled with ethylene glycol. The vessel is held in an adiabatic material (Mica glass-ceramics (Photoveel®)) as shown in Fig. 5. The temperature on both sides of the vessel was controlled by a heater-heat sink system to create a one-dimensional temperature distribution in the liquid medium.

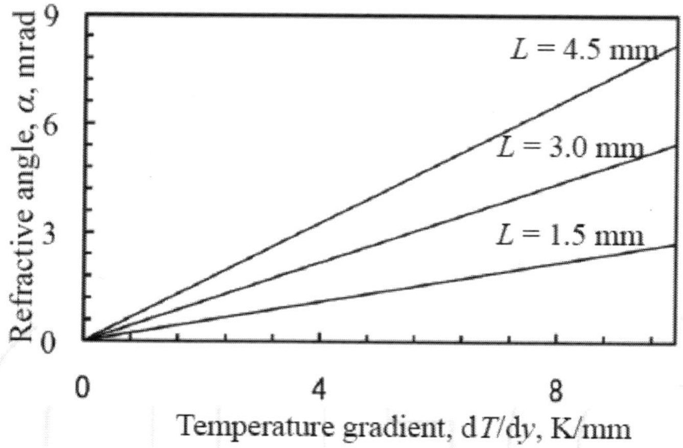

Figure 3. Relationship between the refractive angle and temperature gradient

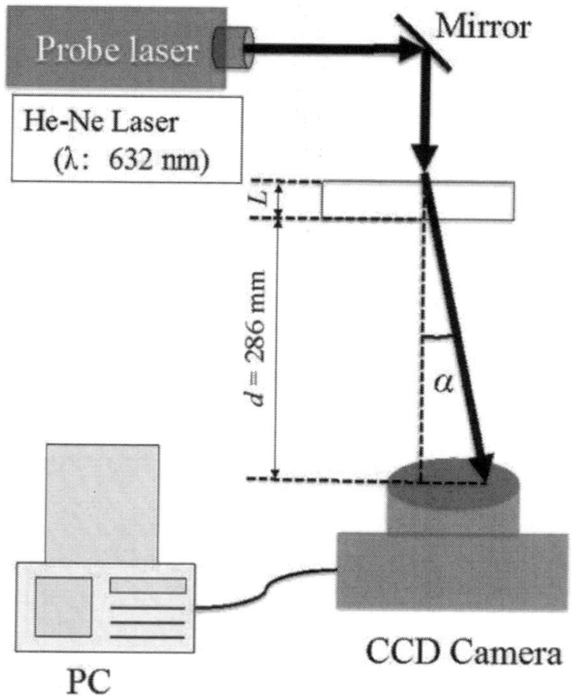

Figure 4. Refractive angle measurement system

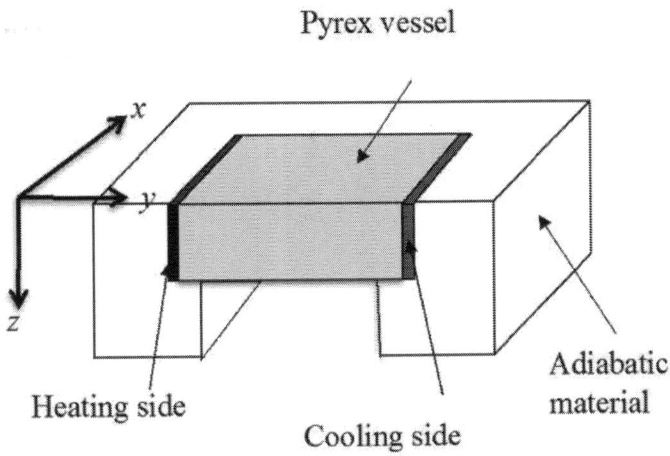

Figure 5. Schematic diagram of sample

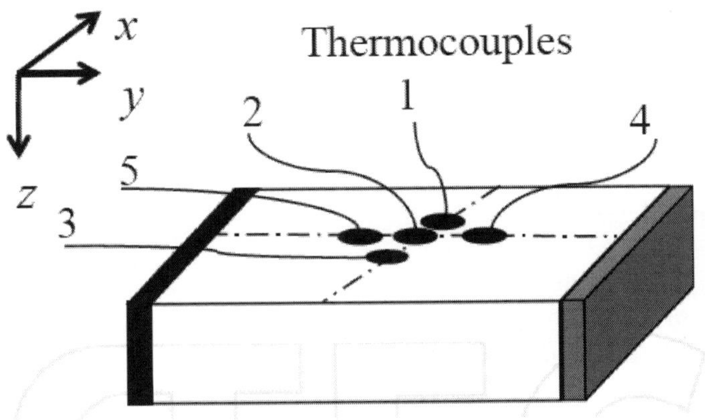

Figure 6. Temperature gradient measurement system

The temperature distribution in the liquid medium is confirmed by measuring the temperatures at 5 points with 2 mm pitch in the vessel using 5 thermocouples as shown in Fig. 6. The temperature gradient in the experiment is given as:

$$\frac{dT}{dy} = \frac{T_4 - T_5}{\Delta y} \tag{11}$$

In which, Δy is the distance between two thermocouples to obtain the temperature gradient, $\Delta y = 4$ mm.

A CW laser ($P = 0.6$ mW, $\lambda = 632$ nm, $\Phi = 0.8$ mm, TEM00) is used as a probe beam. A CCD camera (OPHIR, BeamStar-FX 50) is used as a detector. Based on probe beam position, the refractive angle is estimated with:

$$\alpha = \frac{r}{d} \tag{12}$$

Where d is the distance from the sample to the detector of the camera = 286 mm; r is the beam position.

2.3. Results and Discussions

Figure 7(a) and (b) show the comparison of theoretical and experimental results at points where the thickness of sample, L, is 1.5 and 3.0 mm respectively. The temperature gradient in the theoretical results is based on the measurement as described above. As shown in this figure, theoretical and experimental results agree well with each other. The experimental data includes the error corresponding to the difference between the actual temperature gradient at the laser incidence which is calculated by using Eq. (11). The discrepancy at higher dT/dy may correspond to the error where the measured temperature gives higher dT/dy and therefore higher a prediction by using Eq. (10). This discrepancy should increase with increasing sample thickness and the temperature gradient as a consequence of the effect of natural convection and the temperature gradient in the z-axis [14].

3. FLUIDIC LASER BEAM SHAPER

Flat-top laser are well known to present significant advantages for laser technology, such as holographic recording system, Z-scan measurement, laser heat treatment and surface annealing in microelectronics and various nonlinear optical processes [15-19]. For CW beams, several approaches to spatially shape Gaussian beams have been developed, such as the use of aspheric lenses, implement beam shaping or the use of diffractive optical devices [20]. However, these methods have some disadvantages: a refractive beam shaping system lead to large aberration [21] and implemental beam shaping has low energy efficiency and lacks of flexibility [22]; and the use of refractive optical devices requires complex configuration design and high cost [23]. In practice, a low-cost and flexible method to convert a Gaussian beam into a flat-top beam is required. In this section, a novel method to convert a Gaussian beam into a flat-top beam is discussed. The concept is based on the control of the pump power and propagation distance of the probe beam in the thermal lens system.

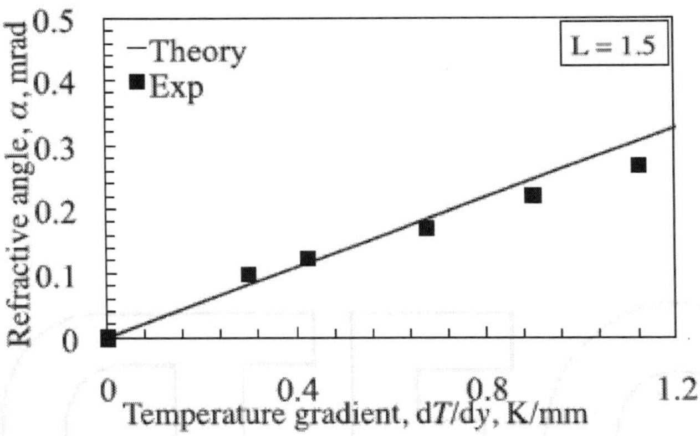

(a) $L = 1.5$ mm

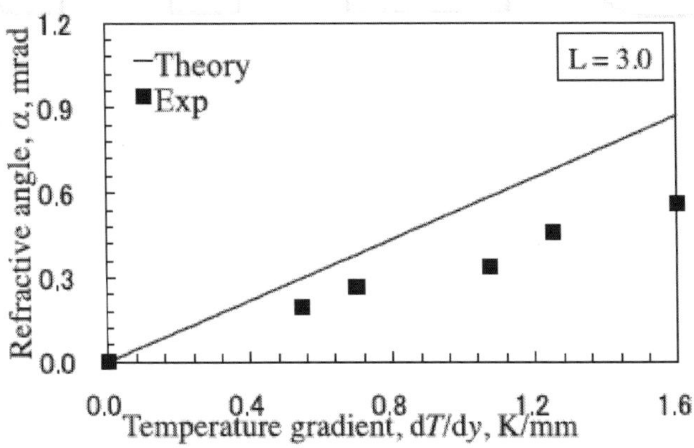

(b) $L = 3$ mm

Figure 7. The comparison of theoretical and experimental results

3.1. Principle Of Thermal Lens Effect

The principle of the transient lens effect is schematically illustrated in Fig.8. A CW diode pumped blue laser is used as pump-beam (BCL-473-030, λ = 473 nm, Φ = 0.8 mm, TEM_{00}) with maximum output power of 30mW. The laser beam intensity was adjusted by using ND-filter. A CW infrared DPSS laser is used as probe-beam (MIL, λ = 1064 nm, Φ = 3.0 mm, TEM_{00}) with maximum output power of 10 mW. A CCD camera is used as a detector to measure the intensity distribution of the laser beam. A cuvette, which is a three-layer structure with a sheet copper is sandwiched between 2 pieces of fused silica. The height of the fused silica is 1 mm. The sheet copper has doughnut shape. The liquid that is contained inside the doughnut hole has the same height with the sheet copper. By varying the thickness of the sheet copper, the liquid height can be changed. The ethanol solution dissolved dye termed as Sunset-yellow is filled in the cuvette. The chemical formula of the Sunset-yellow is shown in Ref. 24.

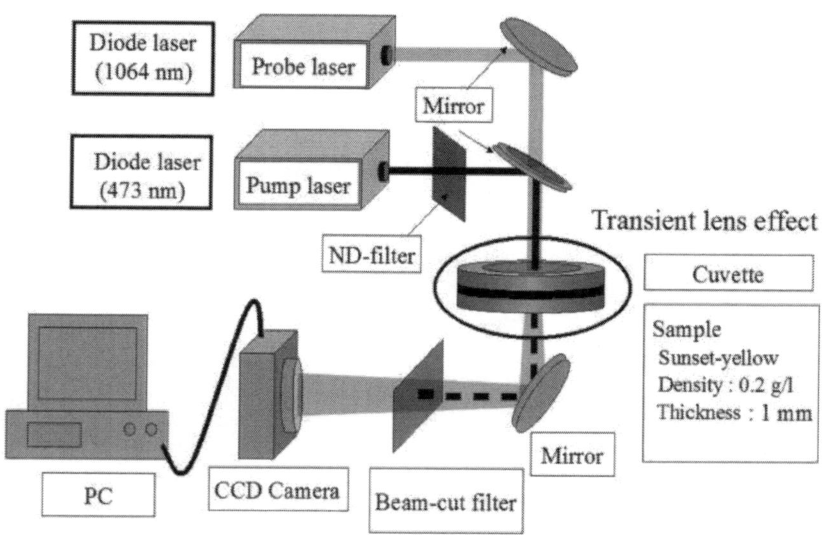

Figure 8. Schematic diagram of a dual-beam thermal lens system

In this experiment, the absorbance of the pump-beam is 2.776 and that of the probe-beam is negligible small. Figure 9 shows the laser beam profile of the probe beam after propagating through a divergent lens and a thermal lens. It is clear that, the probe beam change its profile from Gaussian to doughnut beam with a hollow center is created.

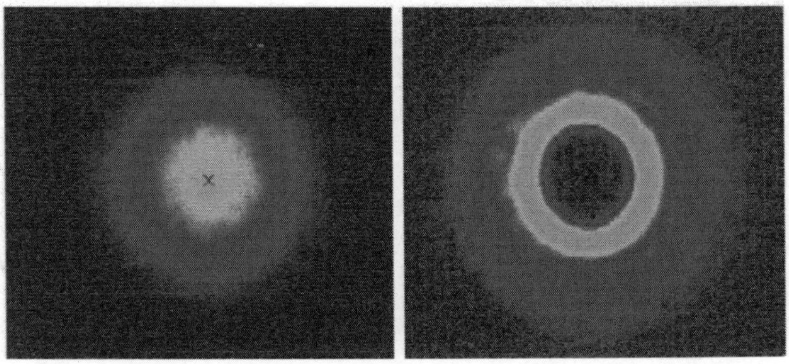

Figure 9. Experiment result: the difference of intensity profile of the probe beam after passing through the thermal lens (left side) and a quart divergent lens (right side)

Theoretical analysis of laser beam profile change in thermal lens effect is done with a model that includes continuity equation, Navier-Stokes equation, energy conservation equation and Helmholtz equation in 2D cylindrical symmetry coordinate. It is assumed that the change of refractive index is caused only by the temperature change of the liquid medium and the thermal coefficient of the refractive index, dn/dT. The concentration is supposed to be constant over the range of the temperature rise induced by the pump beam. When the liquid medium is irradiated, temperature distribution perpendicular to the optical axis is formed due to intensity distribution of laser beam and heat transport. To consider the natural convection effect, the temperature distribution of liquid sample in steady state is calculated numerically following these governing equations [24]:

$$\frac{1}{r}\frac{\partial}{\partial r}\left(rv_r\right)+v_z\frac{\partial v_z}{\partial z}=0 \tag{13}$$

$$v_r\frac{\partial v_r}{\partial r}+v_z\frac{\partial v_r}{\partial z}=-\frac{1}{\rho}\frac{\partial p}{\partial r}+\nu\left(\frac{\partial}{\partial r}\left(\frac{1}{r}\frac{\partial}{\partial r}\left(rv_r\right)\right)+\frac{\partial^2 v_r}{\partial z^2}\right) \quad (a)$$

$$v_r\frac{\partial v_z}{\partial r}+v_z\frac{\partial v_z}{\partial z}=-\frac{1}{\rho}\frac{\partial p}{\partial z}+\nu\left(\frac{1}{r}\frac{\partial}{\partial r}\left(r\frac{\partial v_z}{\partial r}\right)+\frac{\partial^2 v_z}{\partial z^2}\right)+g\beta\left(T-T_0\right) \quad (b)$$

(14)

$$v_r \frac{\partial T}{\partial r} + v_z \frac{\partial T}{\partial z} = a\left(\frac{1}{r}\frac{\partial}{\partial r}\left(r\frac{\partial T}{\partial r}\right) + \frac{\partial^2 T}{\partial z^2}\right) + S \tag{15}$$

$$S = \frac{\alpha e^{-\alpha z} I_0(r)}{\rho C_p} \tag{16}$$

Here, $I_0(r)$ is the intensity distribution of the pump laser. The spot sizes of the laser beams are assumed to be constant through the interaction volume within the liquid medium.

The temperature distribution is calculated numerically based on the finite difference method. The 1st order upwind scheme and a 2nd order center differencing are applied to discretize the advection term and the diffusion term respectively. The thermal properties of liquid medium can be found in Ref. 24.

To model the propagation of laser through an inhomogeneous medium, the wave equation which includes an absorption term and an inhomogeneous refractive index term is applied [24]:

$$-\frac{\partial^2 E}{\partial z^2} + 2ik_0 n_0 \frac{\partial E}{\partial z} = \frac{1}{r}\frac{\partial}{\partial r}\left(\frac{\partial (rE)}{\partial r}\right) + k_0^2\left(n^2 - n_0^2\right)E - \frac{1}{2} ik_0 n_0 \alpha E \tag{17}$$

Here, E is the envelope of the oscillating electric field, z is the axis of propagation, r is transverse coordinates, k_0 is the free space wave number and α is the absorption coefficient. The variable n is the refractive index profile depending on medium temperature following:

$$n(T) = n_0 + \frac{dn}{dT}(T - T_0) \tag{18}$$

Here, $n_0 = 1.359$ is the refractive index of the liquid medium at reference temperature $T_0 = 298.15$ K, dn/dT is the temperature coefficient of the refractive index. The propagation of laser is calculated based on Pade method. The optical properties parameter can be found in Ref. 24.

3.2. Influences Of The Pump Power And The Propagation Distance On The Change Of Probe Beam Profile

Influences of the pump power and the propagation distance to the probe beam profile were investigated numerically using the calculation parameters in Table. 1. In this calculation, both of the pump beam and the probe beam are written as follows.

$$E = E_0 \exp\left(\frac{-r^2}{r_0^2}\right)\exp\left(\frac{-ik_0 n_0 r^2}{2R}\right) \tag{19}$$

$$E_0 = \sqrt{\frac{2P}{\pi r_0^2}} \tag{20}$$

Here *P* is the power of laser, *R* is the radius of curvature of the wave front, *r* and *r0* are distance from laser axis and beam radius respectively.

Table 1. Calculation conditions

Parameter	(a)	(b)
Pump power, mW	3	0 ~ 7
Pump beam diameter, mm	0.8	0.8
Probe power, mW	10	10
Probe beam diameter, mm	0.8	0.8

Absorption coefficient, cm^{-1}	2.0	2.0
Distance from experimental section to CCD camera, mm	0 ~ 500	200
Phase front curvature radius, *R*, mm	∞	∞

Effects of the pump power and the propagation distance to the probe beam profile are shown in Fig. 10(a) and (b) respectively. The vertical axis and horizontal axis show intensity and distance from laser axis respectively. Plots of 'P = 0 mW' and 'd = 0 mm' represent intensity distribution of the probe beam without thermal lens effect. As shown in Fig. 10(a), the further the propagation distance, the lower intensity at the probe beam center, and higher intensity at the wing. With increasing of the propagation distance, the laser beam profile changes from Gaussian to flat-top and the doughnut beam profile respectively. The profile of the probe beam changes with the same tendency as the increasing of pump power as shown in Fig. 10(b). In particular, when the pump power is 3 mW and propagation distance is 200 mm the probe beam is converted to the flat-top profile approximately. Therefore, by controlling the pump power and the propagation distance the Gaussian beam can be converted into the flat-top beam.

3.3. Experimental Set-Up To Shape Spatial Profile

In order to confirm the role of the fluidic laser beam shaper, a single-beam experiment is set up as shown in Fig. 11. A CW diode blue laser is used as pump and probe-beam (P = 10 mW, λ = 488 nm, Φ = 0.69 mm, TEM00). In this experiment, the height of the liquid medium is 0.5 mm, the dye concentration is 0.1 g/l and the absorption coefficient is 2.92 cm^{-1} (measured value) respectively. The propagation distance to obtain the flat-top beam profile is measured by changing the distance from the cuvette to the CCD camera. At the propagation distance of 150 mm, the flat-top beam is confirmed as shown in Fig. 12(a).

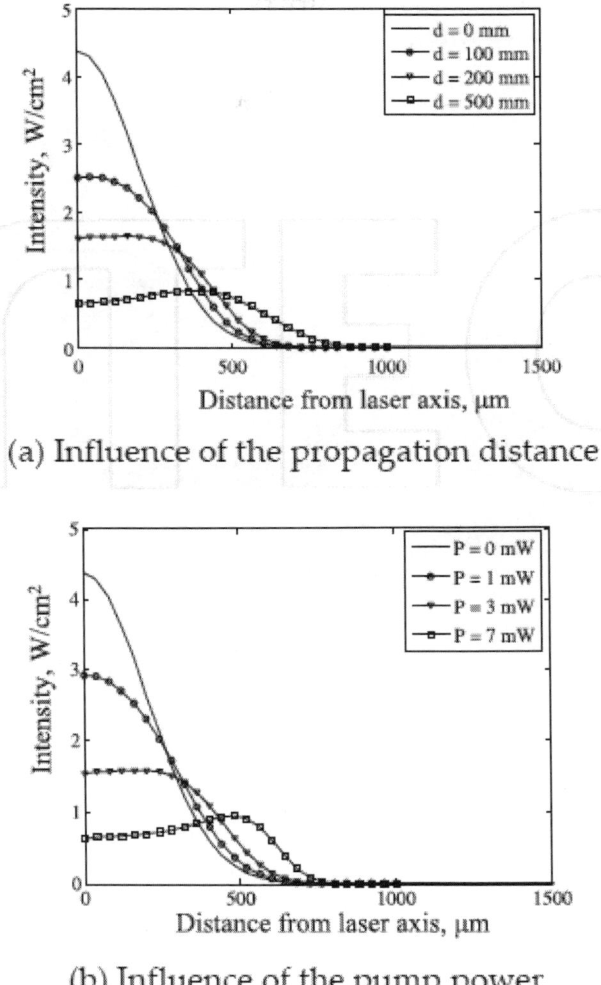

Figure 10. Influence of the propagation distance and the pump power to the probe beam profile

Figure 12(b) shows the beam profile change from the Gaussian to the flat-top beam. The vertical and horizontal axes show the intensity and distance from the laser axis respectively. The o-line shows the profile of the Gaussian input beam by fitting the laser beam profile measured at the surface of the cuvette. The strange-line shows the profile of the flat-top beam calculated by beam propagation method. The solid-line shows the profile of the flat-top beam measured by CCD camera at propagation distance of 150 mm from the cuvette. Both experimental and calculated results agree well with each other.

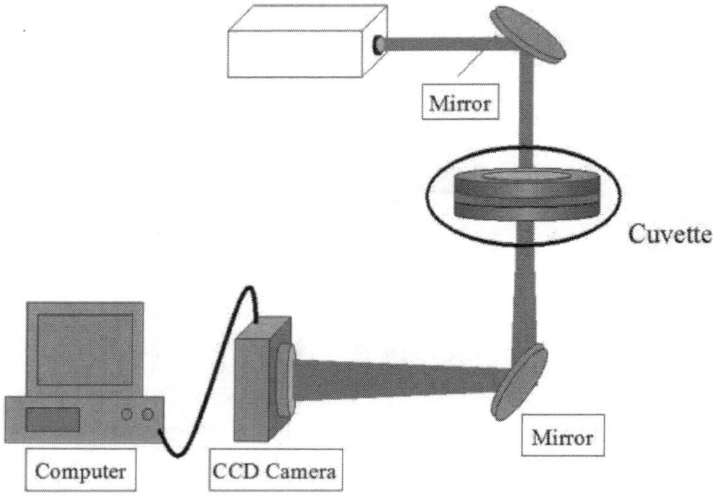

Figure 11. Experimental set up for a single-beam thermal lens system to transfer a Gaussian beam to flat-top beam

In order to explain in more detail about the mechanism of this fluidic beam shaper, the temperature distribution of liquid medium is calculated. As shown in Fig. 13, local heating near the beam axis produces a radially dependent temperature variation, which changes the liquid refractive index in which the lower refractive index is in the region near to the beam center. As a consequence, the radius of curvature of the wave front at the region near the beam center is shorter than one at the beam wing. Therefore the sample liquid locally acts as a micro divergent lens with shorter focal length at beam center. As shown in Fig. 1b, the beam center that passes through shorter focal length is spread out more rapidly than the beam wing. As the probe beam propagates to increasing distance, the intensity in the center region drops rapidly than one in the wing region. At a certain value of propagation distance, the Gaussian beam can be converted into the flat-top beam.

It is noted that, in the case of single-beam shaper, one part of laser beam energy (about 15% in this experiment) is converted into thermal energy in order to change temperature distribution or in other words to change refractive index distribution in the liquid medium. Therefore, in the case of single-beam shaper, the beam shaper has another role, which is as an attenuator. This laser beam shaper/attenuator can be applied in practical laser drilling technology. In the case of applying on only laser beam shaper, the double-beam system is recommended. In this case, it is needed to select dye whose absorbance of the probe-beam is negligible small.

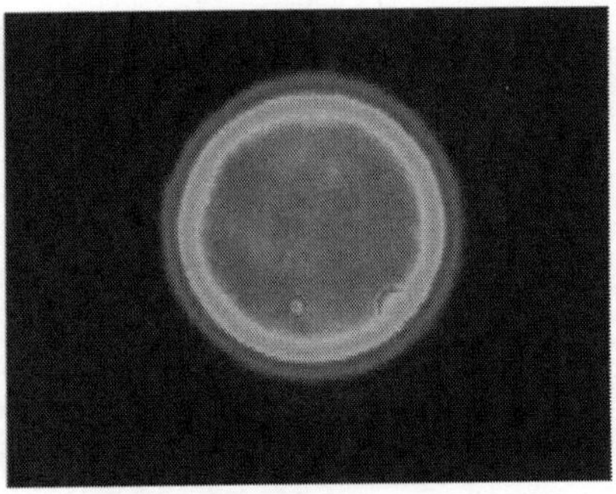

(a) Flat-top beam profile measured by CCD camera

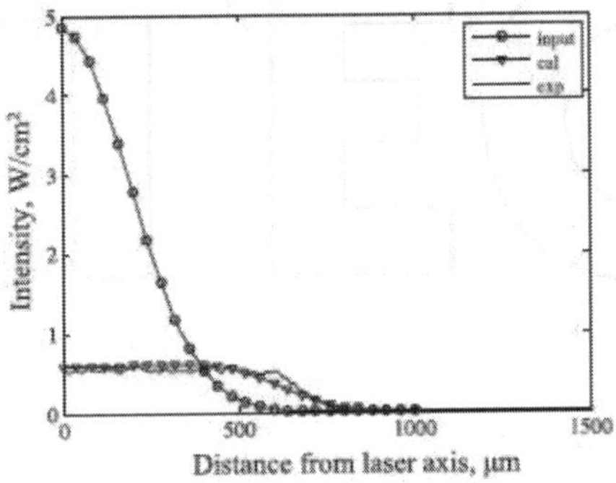

(b) Beam profile change from Gaussian to flat-top

Figure 12. Experimental results

3.4. Relationship Between Pump Power and Distance To Shape Spatial Profile

As shown in previous section, the flat-top beam can be obtained only at a fixed distance. In order to control this distance, the influence of pump power is investigated theoretically and experimentally. The calculation parameters are shown in Table. 2. The pump power is changed from 1 to 8 mW. The distance to obtain the flat-top beam is obtained numerically. The relationship between the pump power and the distance to shape spatial profile is shown in Fig. 14(a). The horizontal and vertical axes show pump power and distance to obtain the flat-top beam respectively. As shown in Fig. 14(a), the distance to obtain a flat-top beam is in inverse proportion to the pump power.

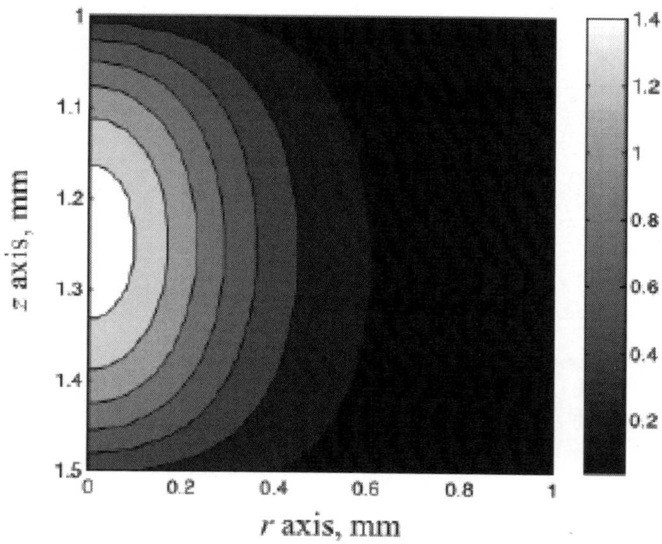

Figure 13. Temperature distribution inside the liquid medium [K]. The calibration shows the difference between temperature inside the liquid medium with the ambient temperature.

In order to validate the numerical prediction, a single beam experiment was carried out. The pump power is changed from 1 to 6 mW and the distance to obtain the flat-top beam was measured. The experimental result shown in Fig. 14(b), shows excellent agreement with calculation prediction. The relationship between pump power and distance to obtain the flat-top beam can be explained by the interaction between energy absorption of liquid medium with the focal length of local micro lens. As the pump power increase, the absorption energy increases. As a consequence, the rate of decreasing of$R$ is enhanced. This can be thought as the reason why the distance to obtain a flat-top beam decreases. In other words, the distance to obtain the flat-top beam profile also decreases with the increasing of absorption coefficient.

Therefore, by changing the absorption coefficient or the pump power, the distance to obtain a flat-top beam can be controlled.

Table 2. Calculation conditions

Pump power, mW	1 ~ 8
Pump beam diameter, mm	0.8
Probe power, mW	10
Probe beam diameter, mm	0.8
Absorption coefficient, cm^{-1}	2.0
Phase front curvature radius, *R*, mm	320

4. TUNABLE FLUIDIC LENS

Fluidic lenses are well known to present significant advantages for wide range of applications from mobile phone to laboratory on a chip. Fluidic lenses have a number of apparent advantages such as tunable refractive index and reconfigurable geometry. Several approaches to design the liquid lens have been developed based on the microfluidic techniques to modify the liquid lens shape by using: out-of-plane micro-optofluidic [25-26], in-plane micro-optofluidic [27-28], electron wetting [29], dielectrophoresis [30] and hydrodynamic force [31]. Other approach bases on turning the refractive index of the liquid by different means such as pressure control, optical control, magnetic control, thermo-optic control, and electro-optic control.

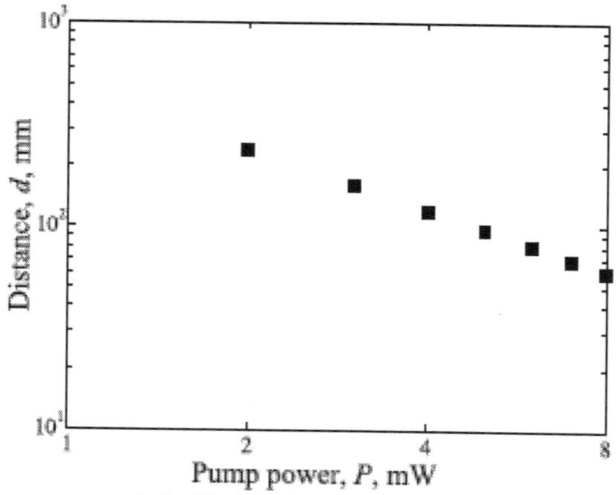

(a) Calculation result

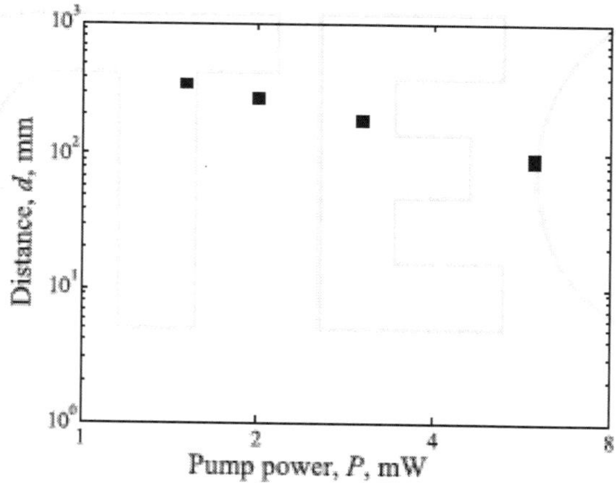

(b) Experimental result

Figure 14. Relationship between the pump power and the distance to obtain the flat-top beam profile. The horizontal and vertical axes show the pump power and the distance to obtain the flat-top beam profile respectively.

4.1. Principle of Fluidic Lens

When the liquid medium is irradiated, local heating near the beam axis produces a radially dependent temperature variation, which changes the liquid refractive index in which the lower refractive index is in the region near to the beam center. As a consequence, the radius of curvature of the wave front at the region near the beam center is shorter than one at the beam wing. The liquid medium behaviors as a convergence GRIN-L with focal length depends on the radial position of the incident ray relative to the optical axis of the cuvette. The ray equation that is calculated numerically to obtain the path of an incident beam, which is given by:

$$\frac{d}{ds}\left(n\frac{dR}{ds}\right)=\text{grad}(n) \tag{21}$$

Where, d*s* and *R* are the differential element of the path length and the positional vector of the ray respectively. The variable *n* is the refractive index of the liquid sample. The variable *n* is the refractive index profile depending on medium temperature following equation (13-16, 18).

4.2. Influences Of The Pump Beam Profile

Influences of the pump beam profile to the focal length of the GRIN-L were investigated numerically with the calculation conditions in Table. 3. The intensity profile of the pump is applied with the Gaussian beam and the quasi-flat-top beam (a super-Gaussian distribution of order *k*) using Eq. 22 andEq. 23 respectively.

$$I_{\text{Gaussian}}=\frac{2P}{\pi r_0^2}\exp\left(\frac{-2r^2}{r_0^2}\right) \tag{22}$$

$$I_{\text{Flat-top}}=\frac{Pk2^{2/k}}{2\pi r_0^2\Gamma(2/k)}\exp\left(\frac{-2r^k}{r_0^k}\right) \tag{23}$$

Here Γ is the Gamma function, *r* and *r0* are distance from laser axis and beam radius, respectively. In this calculation, quasi-flat-top beam is the 10 order of the super-Gaussian distribution, two types of the pump intensity profile are shown in Fig.15.

The effect of the pump beam profile to the focal length of the GRIN-L lens is shown in Fig. 16. The vertical and horizontal axes show focal length and distance from laser axis respectively. The solid and dashed lines represent the plot of the focal length again the radial position of the incident ray relative to the optical axis of the cuvette in the case of

Gaussian pump beam and quasi flat-top pump beam respectively. As shown in Fig. 16, for the Gaussian pump beam the focal length of the GRIN-L increases sharply with increasing of the distance from laser axis, which means larger spherical aberration. It means that, the beam center which passes through shorter focal length is spread out more rapidly than the beam wing. As a consequence, the further the propagation distance of the probe beam, the laser beam profile changes from Gaussian to the doughnut beam profile [24], which should cause some undesirable results in laser processing [32]. In contrast, with the quasi flat-top pump beam, the focal length of the GRIN-L varies lightly with increasing of the distance from laser axis smaller than beam waist of the flat-top pump beam. The area smaller than the beam waist of the flat-top pump beam acts as a divergent lens with small spherical aberration. Therefore, for the purpose of designing the GRIN-L lens the uniform pump beam shows the advance in reducing the spherical aberration.

Table 3. Calculation conditions

Pump power, mW	10
Pump beam diameter, mm	1.5
Absorption coefficient, cm^{-1}	2.0

4.3. Experimental Set-Up

In order to confirm the qualities of the GRIN-L, an experiment with the quasi flat-top pump beam is carried out as shown in Fig. 17. A CW diode blue laser is used as pump laser (P = 10 mW, λ= 488 nm,Φ = 0.69 mm, TEM00). In cuvette 1, the height of liquid is 0.5 mm, and the absorption coefficient is 2.92 cm^{-1} (at wavelength of 488 nm). In the cuvette 2, the height of liquid is 1 mm, and the absorption coefficient is 55 cm^{-1} (at wavelength of 488 nm). A CW He-Ne laser is used as probe laser (P = 0.6 mW, λ = 632 nm, Φ = 0.8 mm, TEM00). It is noted that, the absorption of ethanol solution can be ignored at the wavelength of the probe laser.

First, the pump beam passes through cuvette1, then the beam profile of pump beam was converted from Gaussian to flat-top during its transmission to cuvette 2 as shown in Fig. 18. Then, the probe laser was adjusted to overlap with pump laser. After propagating through the sample the probe laser is directed towards the CCD camera and the pump laser is blocked using filters located at the detection plane. The distance between cuvette 2 and the CCD camera is varied, and the $1/e^2$ diameter of probe laser is measured.

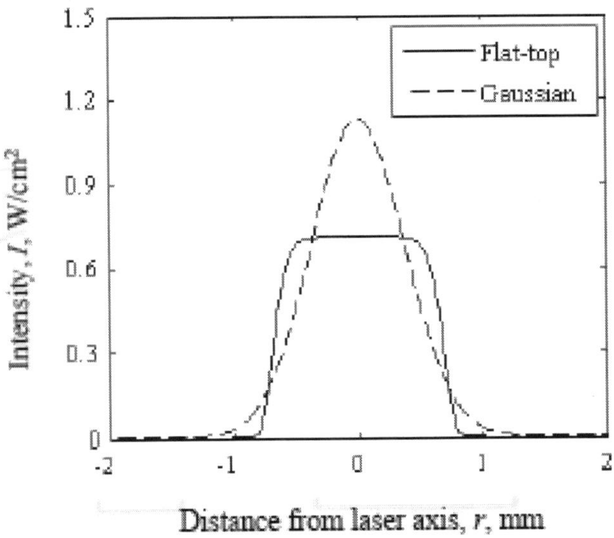

Figure 15. Two types of the pump beam profile

Figure 19(a) shows the change along the propagation direction in the beam profile. The vertical and horizontal axes show the intensity and distance from the laser axis respectively. By using the quasi flat-top pump beam, the beam profile of probe laser can remain in Gaussian distribution during its propagation. Figure 19(b) shows the plot of probe beam waist again propagation distance. As shown in Fig. 19(b), the beam waist of probe laser varies linearly with propagation distance. In other words, cuvette 2 acts as a divergence lens with focal length of f = -424 mm (this value has been calculated by considering the divergence angle of probe laser θ = 1.2 mrad).

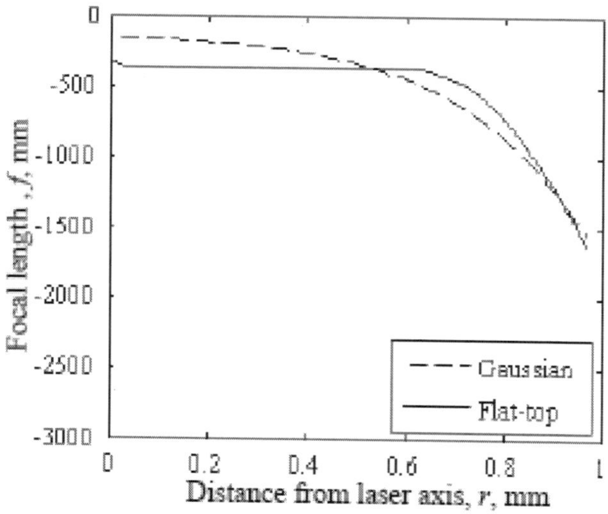

Figure 16. Effect of the pump beam profile to the focal length of the GRIN-L lens

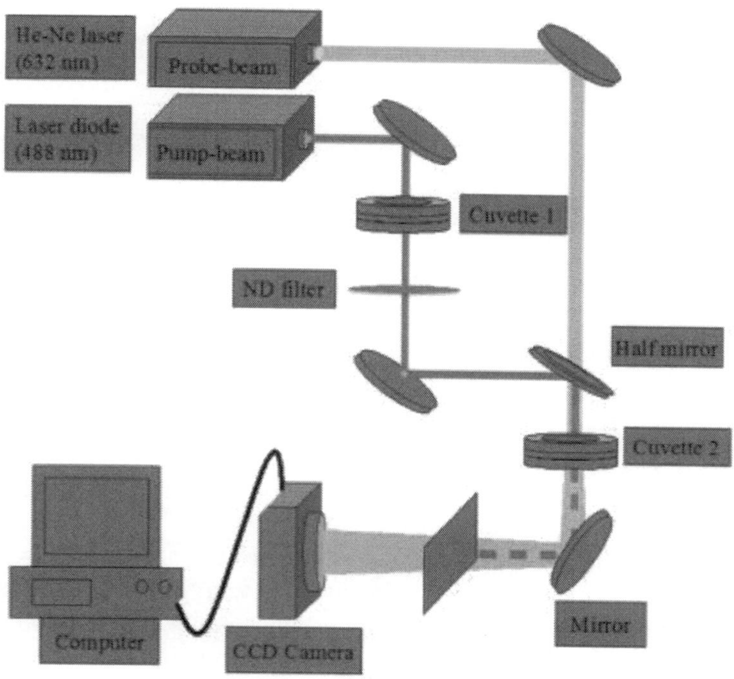

Figure 17. Experimental set-up for the fluidic divergent lens

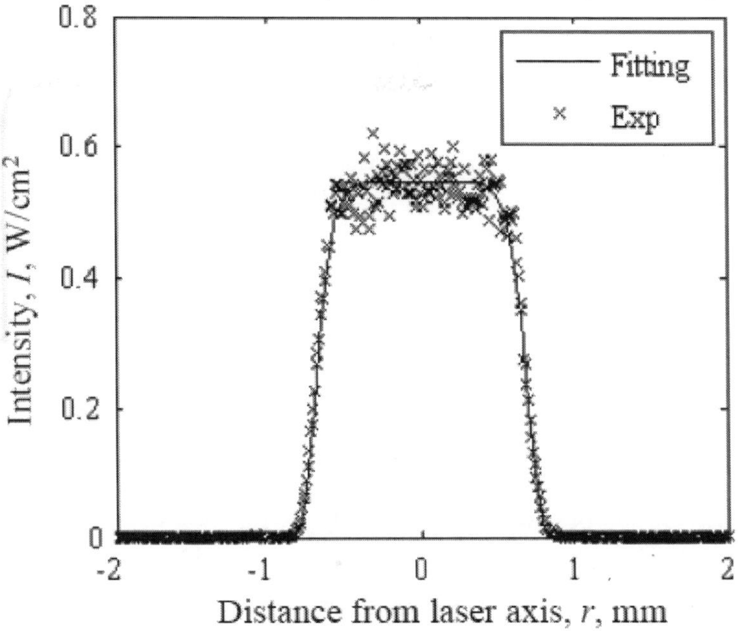

Figure 18. The intensity profile of the pump beam during its transmission to cuvette 2. Dotted and solid lines show the measured result and fitting by super-Gaussian distribution respectively.

Next, the pump power is changed from P_0 = 7.7 mW to $P_0/2$, $P_0/3$ and $P_0/4$ respectively. Figure 20shows the plots of focal length against the pump power. Square and circle plots show the calculation and experimental result, respectively. As shown in Fig. 20, the focal length increases with increasing of the pump power. This means that, by adjusting the pump power, the focal length can be controlled.

5. CONCLUSION

In this research, a novel idea of fluidic optical devices which includes laser beam shaper and fluidic divergent lens are demonstrated. The fluidic optical devices are based on controlling some parameters in the thermal lens system. The interaction among the intensity distribution, power of the pump beam, the absorption coefficient, the propagation distance and the intensity profile of the probe beam have been investigated experimentally and theoretically. It is found that

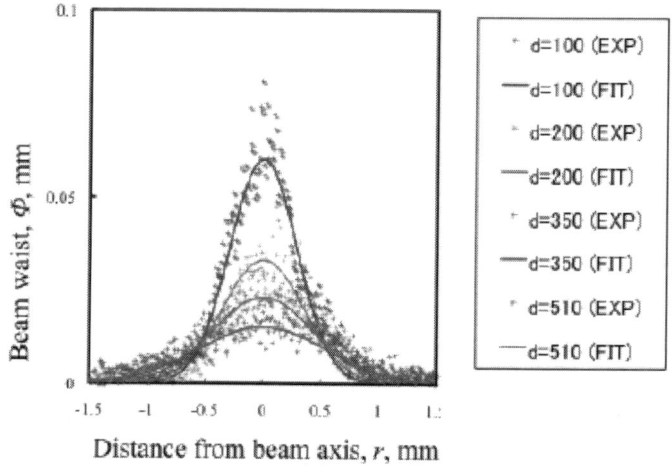

(a) Beam profile

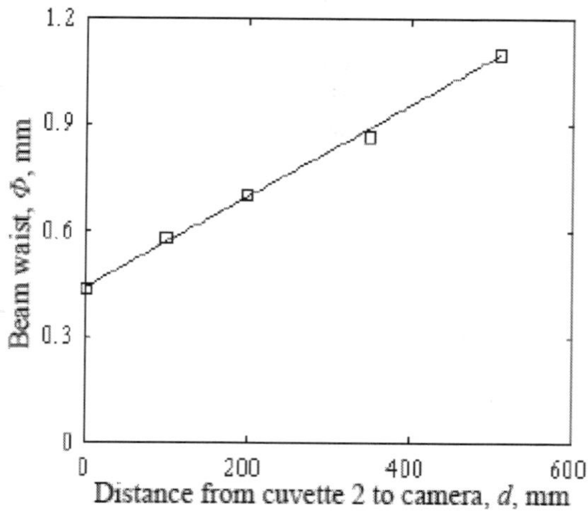

(b) Beam waist

Figure 19. Probe beam changes along the propagation direction

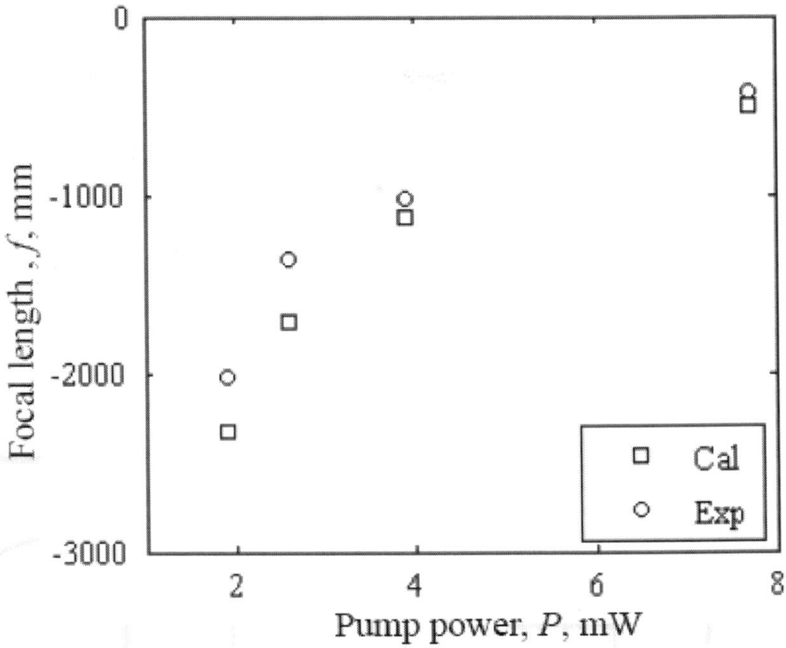

Figure 20. Relationship between the pump power and focal length

- By controlling the pump power and the absorption coefficient, the input Gaussian beam can be converted into a flat-top beam profile. The distance to get the flat-top beam profile can be controlled easily by adjusting the pump power and the absorption coefficient. In actual applications, single-beam shaper has another role, which is as an attenuator. This laser beam shaper/attenuator can be applied in practical laser drilling technology. In the case of applying on only laser beam shaper, the double-beam system is recommended. In this case, it is needed to select a dye whose absorbance of the probe-beam is negligible small.

- The uniform pump beam shows the advance in reducing the spherical aberration. And by adjusting the pump power, the focal length can be controlled

With some merits such as flexiblility, versatility and low cost, these fluidic optical devices will be promising tools in many fields of laser application.

ACKNOWLEDGEMENT

Part of this work has been supported by the Grant-in-Aid for JSPS Fellows and Grant-in-Aid for Scientific Research of MEXT/JSPS. The authors also would like to acknowledge Mr. Akamine Yoshihiko.

REFERENCES

1. J. P. Gordon, R. C. C. Leite, R. S. Moore, S. P. S. Porto, J. R. Whinnery, . Long-Transient, in. effects, with. lasers, liquid. inserted, J. samples, Appl. Phys. 36 (196538

2. S. A. Akhmanov, D. P. Krindach, A. V. Migulin, A. P. Sukhorukov, R. V. Khokhlov, self-actions. Thermal, laser. of, I. E. E. E. J. beams, Quantum Electronics QE-419681968568

3. P. M. Livingston, induced. Thermally, of. a. modifications, power. C. W. high, beam. laser, Appl, Opt. 1019711971426436

4. J. F. Power, mode. Pulsed, lens. thermal, detection. effect, the. in, field. near, thermally. via, probe. induced, spatial. beam, modulation. a. phase, Appl. theory, Opt. 291990199052

5. P. P. Banerjee, R. M. Misra, M. Maghraoui, Theoretical, studies. experimental, propagation. of, beams. of, a. through, sample. finite, a. of, nonlinear. cubically, J. material, Opt. Soc. Am. B 819911991107210 80

6. J. M. Hickmann, A. S. L. Gomes, C. B. de Araújo, of. Observation, cross-phase. spatial, effects. modulation, a. in, nonlinear. self-defocusing, Phys. medium, Rev. Lett. 681992199235473550

7. P. Govind, Transverse. Agrawal, instability. modulation, copropagating. of, beams. optical, nonlinear. in, media. J. Kerr, Opt. Soc. Am. B 71990199010721078

8. C. J. Rosenberg, et al.of. Analysis, dynamics. the, high. of, Gaussian. intensity, beams. laser, nonlinear. in, Kerr. de-focusing, Optics. media, 2. Communications, 2007

9. M. Sakakura, M. Terazima, of. Oscillation, refractive. the, at. index, focal. the, of. a. region, laser. femtosecond, inside. a. pulse, Opt. glass, lett. 29 (13) (2004

10. M. Sakakura, M. Terazima, observation. Real-time, photothermal. of, after. effect, of. photo-irradiation, laser. femtosecond, inside. a. pulse, J. glass, Phys. France 12520052005355360

11. M. Terazima, N. Hirota, S. E. Braslavsky, A. Mandelis, S. E. Bialkowski, G. J. Diebold, R. J. D. Miller, D. Fournier, R. A. Palmer, A. Tam, terminology. Quantities, in. symbols, photothermal, spectroscopies. . I. U. P. A. C. related, Recommendations, 2004Pure Appl.Chem., 76, 1083

12. Kudou Uehara1990Basic Optics (Kougaku Kiso) Gendaikougaku, Tokyo, Japan, 4547Japanese)

13. S. K. Y. Tang, B. T. Mayers, D. V. Vezenov, G. M. Whitesides, waveguiding. Optical, thermal. using, across. gradients, liquids. homogeneous, microfluidic. in, Appl. channels, Phys. Lett. 88, 06112, 2006

14. H. D. Doan, K. Fushinobu, K. Okazaki, on. Investigation, interaction. the, light. among, material, field. temperature, the. in, lens. transient, transmission. effect, in. . D. characteristics, field. temperature, I. Proc, ITherm 2010127

15. J. Yang, Y. Wang, X. Zhang, C. Li, X. Jin, M. Shui, Y. Song, of. Characterization, transient. the, effect. thermal-lens, flat-top. using, Z-scan. J. beam, Phys. B: At. Mol. Opt. Phys. 42 (2009pp)

16. K. Ebata, K. Fuse, T. Hirai, K. Kurisu, laser. Advanced, for. optics, material. laser, Proc. S. P. I. processing, SPIE, 50632003

17. E. B. S. Govil, J. P. Longtin, A. Gouldstone, M. D. Frame, visible. Uniform-intensity, source. light, in. for, imaging. situ, of. Journal, Optics. Biomedical, 14(, 2009

18. F. M. Dickey, S. C. Holswade, D. L. , Shealy 'Laser Beam Shaping Applications', Taylor & Francis, 2006

19. M. T. Eismann, A. M. Tai, J. N. Cederquist, design. Iterative, a. of, beam. holographic, Appl. former, Opt. 2819981998264111650

20. F. M. Dickey, S. C. Holswade, beam. Laser, Theory. shaping, Marcel. Techniques, New. Dekker, York, 2000

21. P. Scott, optics. Reflective, irradiance. for, of. redistribution, beam. laser, Appl. design, Opt. 20 (9) (1981

22. S. Zhang, Q. Zhang, G. Lupke, beam. Spatial, of. shaping, laser. ultrashort, theory. pulse, Appl. experiment, Opt.442005200 558185823

23. B. Mercier, J. P. Rousseau, A. Jullien, L. Antonucci, beam. Nonlinear, for. shaper, laser. femtosecond, from. pulses, to. Gaussian, profile. flat-top, Communications. 2. Optics, 2010

24. H. D. Doan, Y. Akamine, K. Fushinobu, laser. Fluidic, shaper. beam, using. by, lens. thermal, Int. effect, J. Heat Mass Transfer (2012

25. S. H. Ahn, Y. K. Kim, of. Proposal, eye's. human, lens-like. crystalline, focusing. variable, Sens. lens, Actuators (1999A 78, 48

26. D. Y. Zhang, V. Lien, Y. Berdichevsky, J. H. Choi, Y. H. Lo, adaptive. Fluidic, with. lens, focal. high, tenability. length, Appl, Phys. Lett. 82 (2003

27. S. K. Hsiung, C. H. Lee, G. B. Lee, electrophoresis. Microcapillary, utilizing. chips, micro-lens. controllable, structures, optical. buried, for. fibers, optical. on-line, Electrophoresis. detection, 2008

28. V. Lien, Y. Berdichevsky, Y. H. Lo, surfaces. Microspherical, predefined. with, lengths. focal, using. fabricated, capillaries. microfluidic, Appl, Phys. Lett. (2003

29. C. B. Gorman, H. A. Biebuyck, G. M. Whitesides, of. Control, Shape. the, Liquid. of, on. a. Lenses, Gold. Modified, Using. Surface, Applied.

an, Potential. Electrical, a. across, Monolayer. Self-Assembled, Langmuir, 1995

30. C. C. Cheng, C. A. Chang, H. A. Yeh, focus. Variable, liquid. dielectric, lens. droplet, Opt, Express (2006

31. S. K. Y. Tang, C. A. Stan, G. M. Whitesides, reconfigurable. Dynamically, liquid-cladding. liquid-core, in. a. lens, channel. microfluidic, Chip. Lab, 2008

32. D. H. Doan, Y. Yin, N. Iwatani, K. Fushinobu, processing. Laser, using. by, laser. fluidic, shaper. beam, Proc, National Heat Transfer Symposium 2012

9

DOUBLE-SLOT HYBRID PLASMONIC RING RESONATOR USED FOR OPTICAL SENSORS AND MODULATORS

Xu Sun [1,2,*], Daoxin Dai [2,3], Lars Thylén [1,2,4,5] and Lech Wosinski [1,2]

[1] School of Information and Communication Technology, KTH Royal Institute of Technology, 164 40, Kista, Sweden

[2] JORCEP, Joint Research Center of Photonics of the Royal Institute of Technology (Sweden) and Zhejiang University (China), Hangzhou, 310058, China

[3] Centre for Optical and Electromagnetic research, State Key Laboratory for Modern Optical Instrumentation, Zhejiang University, Zijingang Campus, Hangzhou, 310058, China

[4] Department of Theoretical Chemistry and Biology, KTH Royal Institute of Technology, 106 91 Stockholm, Sweden

[5] Hewlett-Packard Laboratories, Palo Alto, CA 94304, USA

ABSTRACT

An ultra-high sensitivity double-slot hybrid plasmonic (DSHP) ring resonator, used for optical sensors and modulators, is developed. Due to high index contrast, as well as plasmonic enhancement, a considerable part of the optical energy is concentrated in the narrow slots between Si and plasmonic materials (silver is used in this paper), which leads to high sensitivity to the infiltrating materials. By partial opening of the outer plasmonic circular sheet of the DSHP ring, a conventional side-coupled silicon on insulator (SOI) bus waveguide can be used. Experimental results demonstrate ultra-high sensitivity (687.5 nm/RIU) of the developed DSHP ring resonator, which is about five-times higher than for the conventional Si ring with the same geometry. Further discussions show that a very low detection limit (5.37×10^{-6} RIU) can be achieved after loaded Q factor modifications. In addition, the plasmonic metal structures offer also the way to process optical and electronic signals along the same hybrid plasmonic circuits with small capacitance (~0.275 fF) and large electric field, which leads to possible applications in compact high-efficiency electro-optic modulators, where no extra electrodes for electronic signals are required.

KEYWORDS

integrated optics; waveguides; plasmonic

1. INTRODUCTION

The possibilities of miniaturization, mass production, low power consumption and wide bandwidth cause photonic integrated circuits (PICs) [1] to play an increasingly important role in various applications. In particular, for electro-optic (EO) polymer-based optical modulators and refractive index (homogenous) sensors, the waveguide sensitivity to refractive index changes is one of the vital parameters. Many waveguide structures, including dielectric waveguides [2,3,4,5,6,7, 8,9,10,11,12], subwavelength grating waveguides [13], hollow core waveguides [14], dielectric slot waveguides [15,16,17,18,19] and plasmonic slot waveguides [20], have been designed and optimized to increase the optical confinement factor of covering or infiltrating materials and, hence, to increase sensitivity. In contrast to dielectric- and subwavelength grating waveguide-based sensors (evanescent optical sensing), slot waveguides or hollow core waveguides can confine optical power in the narrow slots or hollow cores, which leads to large optical sensitivity to materials infiltrating the waveguides. In the following, we have limited our investigations to slot waveguides. The principles of slot-guiding waveguides are different from ordinary index guiding: for a dielectric slot waveguide, the optical confinement is provided by the special high-low-high indices' structures, where the large index-contrast makes the optical power propagate in the low index slot; for the plasmonic slot waveguide, confinement is achieved by plasmonic optical enhancement, where optical energy is excited and propagated at the surface of metal layers. Generally speaking, plasmonic slot waveguides have better optical confinement than dielectric slot waveguides, due to the ultra-compact size, however, at the expense of larger propagation loss. To combine the advantages of both dielectric and plasmonic slot waveguides, a double-slot hybrid plasmonic (DSHP) waveguide, a novel high optical confinement, low propagation loss waveguide, has been developed [21,22].

The DSHP waveguide structure shows the advantages of low propagation loss, high optical confinement and a fabrication process compatible with conventional silicon on insulator (SOI) technologies. In [22], a DSHP Mach-Zehnder Interferometer (MZI) resonant sensor is experimentally demonstrated. In this paper, we continue our work on the investigation of the DSHP waveguides using here ring resonator structures. In comparison to MZI resonant cavities, the ring resonator has a better quality factor, which will give a smaller detection limit for optical sensor applications and can also decrease the power consumption for electro-optic (EO) modulators.

In this paper, based on detailed simulation investigations, a Si bus waveguide side-coupled DSHP ring is designed, fabricated and characterized, showing the potential of applications in highly-efficient optical homogenous sensors and in EO polymer-based modulators.

2. DOUBLE-SLOT HYBRID PLASMONIC RING

2.1. Schematic and Simulation Method

As shown in Figure 1a, the DSHP ring resonator consists of a Si ring located between Ag (or other plasmonic material) circular sheets. When the slots between the Ag sheets and Si ring are narrow enough (less than 500 nm), the quasi-transverse electric (TE) hybrid photonics (Si-slot materials) and plasmonic (slot material-Ag) guided mode can be supported. In the ring structure, it is propagating as the known whispering-gallery mode.

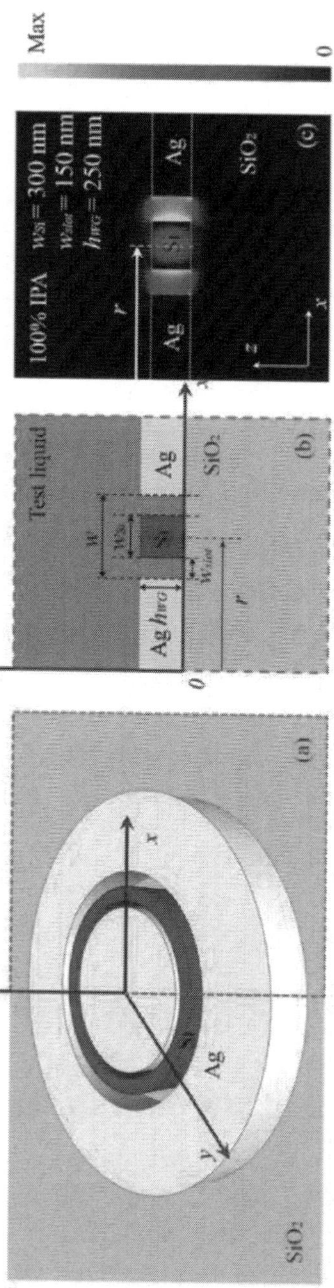

Figure 1. (**a**) Schematic of a double-slot hybrid plasmonic ring resonator; (**b**) x-z plane cross-section view; (**c**) power distribution of the x-z plane cross-section simulated by the axisymmetric finite element method. IPA, 2-isopropanol.

The x-z cross-section view is shown in Figure 1b. SiO_2 is used as the buffer layer, and liquids to be tested (or electro-optic polymer) are filling the slots and covering the device (infiltrating the device). The radius of the DSHP ring r is equal to the middle radius of the Si ring, as shown in Figure 1b. The width of the Si ring is denoted by w_{Si}; the heights of the Si and metal layers are identical and denoted by h_{WG}; the widths of the slots are w_{slot}, and w is the total width ($w = w_{Si} + 2w_{slot}$).

The power distribution of the simulated DSHP ring is shown in Figure 1c using the axisymmetric finite element method (FEM) [23], where the commercial simulation software, COMSOL Multiphysics, is used to solve the partial differential equation in cylindrical coordinates. The geometrical parameters are: h_{WG} = 250 nm, w_{Si} = 300 nm and w_{slot} = 150 nm. The refractive indices of Si and SiO_2 are 3.45 and 1.45, respectively, for the operation wavelength of 1550 nm. The infiltrating material is pure 2-isopropanol (IPA), whose refractive index is 1.3739 [24] at 1550 nm.

The properties of Ag are calculated by the Drude model [25]:

$$\varepsilon = \varepsilon_{\infty} - \frac{\omega_p^2}{\omega^2 + j\omega\gamma} \tag{1}$$

where $\varepsilon_{\infty} = 3.1$, $\omega_p = 140 \times 10^{14}$ rad/s and $\gamma = 0.31 \times 10^{14}$ rad/s.

2.2. Model Investigation

The model investigations start from the generalized analysis of the quality factors (Q factors) of the DSHP ring with various widths of the slots, as shown in Figure 2a. The geometrical parameters of the DSHP ring are: w_{Si} = 350 nm and h_{WG} = 250 nm. In the simulation process, the resonant wavelength is fixed around 1550 nm, and the azimuthal numbers (m) are changed from 15 to 55 with a step of five. The radii of the DSHP ring are adjusted according to the resonant condition of whispering-gallery mode:

$$2\pi r = m\lambda / n_{eff} \tag{2}$$

The Q factor is calculated by:

$$Q = \frac{f_{eig}(real)}{2f_{eig}(imag)} \quad (3)$$

where $f_{eig}(real)$ and $f_{eig}(imag)$ are the real and imaginary parts of the eigenvalue of the *m*-th order whispering-gallery mode, which can be directly readout from the simulation software.

Without considering the coupling loss, the *Q* factor of the DSHP ring is influenced by absorption and radiation losses [26], which can be written as:

$$\frac{1}{Q} = \frac{1}{Q_{abs}} + \frac{1}{Q_{rad}} \quad (4)$$

Here, Q_{abs} and Q_{rad} are the absorption and radiation quality factors of the ring, respectively.

As shown in the figure, the *Q* factor increases with the radius due to lower radiation loss; when the radius is large enough (larger than 6 μm), the radiation loss can be ignored, while the initial properties of the DSHP waveguide (absorption loss) play a more significant role in the *Q* factor, which tends to a constant value for specific geometrical parameters. In addition, the *Q* factor of the DSHP ring with wider slots is larger, which means that the absorption loss of the DSHP waveguide with wider slots is smaller due to lower plasmonic material influence.

Summarizing the information given above, the absorption loss (or Q_{abs} factor) is the major property of a DSHP ring with a large enough radius. In order to study the Q_{abs} factor of the DSHP ring, one can fix the radius of the DSHP ring at 6 μm (radiation loss can be ignored) and change the geometry of the DSHP waveguide. In this way, the normal FEM simulation method can be used to study the properties of a DSHP waveguide *per se*. Since the hybrid propagation mode is a mixture between photonic and Surface Plasmon Polariton (SPP) modes, a *q*-parameter ($q = w_{Si}/w$), representing the occupancy of a photonic mode, is used to study and optimize the performance of the DSHP ring resonator.

Figure 2b shows the Q_{abs} factor *versus* the *q*-parameter, when the total width of the DSHP ring is changed from 500 nm to 1000 nm. The Q_{abs} factor is estimated by:

$$Q_{abs} = \frac{\omega n_g}{c\alpha} \quad (5)$$

where n_g is the group index of the confined whispering-gallery mode, which is expressed by:

$$n_g = n_{eff}(real) - \lambda \frac{dn_{eff}(real)}{d\lambda} \quad (6)$$

$n_{eff}(real)$ is the real part of effective refractive index. The dispersion of n_{eff}, $dn_{eff}(real)/d\lambda$, is computed by re-simulating the DSHP waveguide with a wavelength of $\lambda + \Delta\lambda$, and the n_{eff} at $\lambda + \Delta\lambda$ can be obtained, which is denoted as $n_{eff}(\lambda + \Delta\lambda)$.

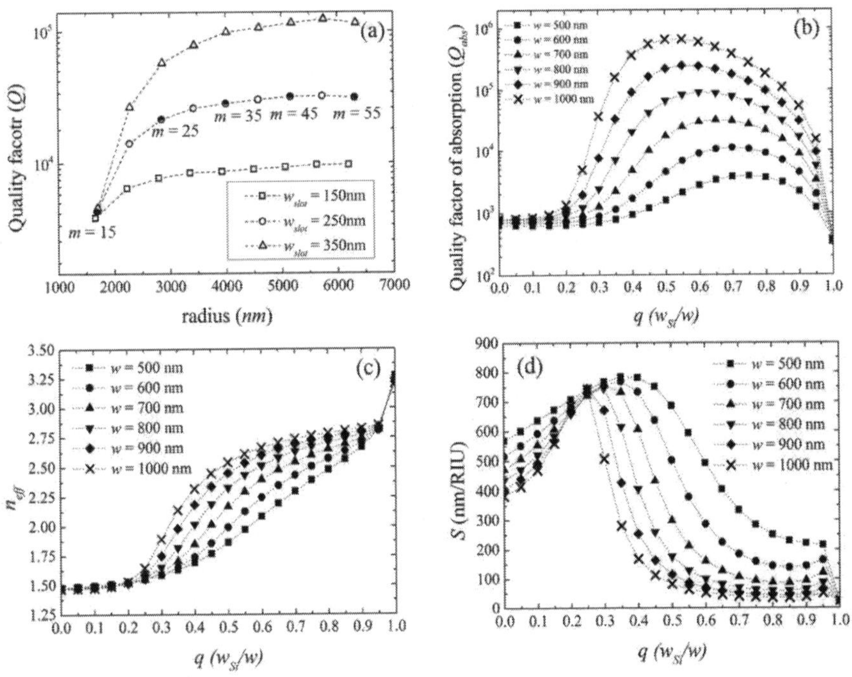

Figure 2. (**a**) *Q* factor *versus* the radius of the double-slot hybrid plasmonic (DSHP) ring resonator with various widths of the slots (150 nm, 250 nm and 350 nm); the other geometrical parameters are: w_{Si} = 350 nm and h_{WG} = 250 nm; (**b,c,d**) the Q_{abs} factor, effective refractive index and sensitivity changes with the *q*-parameter for waveguides with a total width *w* of 500 nm, 600 nm, 700 nm, 800 nm, 900 nm and 1000 nm.

Then, the dispersion can be estimated by:

$$\frac{dn_{eff}(real)}{d\lambda} = \frac{n_{eff}(\lambda+\Delta\lambda) - n_{eff}(\lambda)}{\Delta\lambda} \quad (7)$$

α in Equation (5) is the absorption coefficient, which can be written as:

$$\alpha = \frac{4\pi}{\lambda} \cdot n_{eff}(imag) \quad (8)$$

where $n_{eff}(imag)$ is the imaginary part of effective refractive index.

After simulating the DSHP waveguide, the Q_{abs} factors *versus* the q-parameter are calculated according to Equation (5), as shown in Figure 2b, where the total widths of the DSHP waveguide are changed from 500 nm to 1000 nm with a step of 100 nm. One can observe that the Q_{abs} factor of the DSHP ring decreases with a smaller width due to the larger influence of the plasmonic layers in the narrower DSHP waveguide. For the DSHP rings with the same total width, there are optimal values of the q-parameter giving the maximum Q_{abs} factor, which expresses the hybrid propagation modes of the DSHP ring, where the propagation loss is much lower than that of a pure plasmonic slot waveguide (when q is equal to either zero or one).

Values of n_{eff} *versus* the q-parameter for different total widths are shown in Figure 2c. The n_{eff} increases with the occupancy of the slot by the Si ring (increase of q), which needs to be explained from two aspects: (1) the high refractive index of Si has greater influence and causes higher n_{eff} of the DSHP waveguide; and (2) with the higher q-parameter, w_{slot} of the DSHP ring is small, and so, the optical confinement is larger. The former explanation concerns the photonic mode, and the latter one concerns the plasmonic optical enhancement. The interaction between photonic and plasmonic modes leads to the nonlinear behavior of n_{eff}.

In addition to the basic studies of the DSHP ring, we also analyze its sensitivity to infiltrating materials. The sensitivity is given by:

$$S = \frac{\partial\lambda_{res}}{\partial n_{cover}} = \lambda\frac{\Delta n_{eff}}{n_g} / \partial n_{cover} \quad (9)$$

where $\partial\lambda_{res}$ is the resonant wavelength shift and ∂n_{cover} is the refractive index change of infiltrating materials (∂n_{cover} is set to be 0.003 in the simulation); Δn_{eff} is the change of effective indices of the DSHP waveguide infiltrated with different materials.

As shown in Figure 2d, the sensitivity of the narrower waveguide is higher due to the better optical confinement. In addition, optimal q-parameters are also observed for the sensitivities of the DSHP ring, which means that the sensitivity is also enhanced by the hybrid modes compared to pure plasmonic modes (q = 0 or 1). The maximum sensitivity can reach a value as high as 700 nm/RIU when w = 500 nm and q = 0.35, which can be further increased by decreasing the total width, however, at the expense of a lower Q_{abs} factor. The trade-off between the Q_{abs} factor and sensitivity is similar to the one between propagation loss and optical confinement, which is a general behavior in hybrid plasmonic and plasmonic waveguides [21,27]. Based on the different behaviors of the Q_{abs} factor and the sensitivity of the DSHP ring, one needs to carefully design the geometry of the DSHP ring to satisfy the requirements for particular applications.

3. EXPERIMENTAL REALIZATION

3.1. SI SIDE-COUPLED DSHP RING RESONATOR DESIGN

In a conventional SOI technology, a side-coupled Si bus waveguide is commonly used to characterize the performance of ring resonators. However, in our case, due to the presence of the outer Ag ring, almost no optical energy from the bus waveguide would be coupled into the DSHP ring. To solve this problem, one can reconfigure the DSHP ring with a partly open area, as shown in the sub-figure of the coupling area in Figure 3, where a part of the outer Ag ring is taken away. The gap between the Si bus waveguide and the DSHP waveguide is designed to be 100 nm to 200 nm, to find the experimental critical coupling condition. The geometrical parameters of the DSHP ring are: w_{slot1} = 250 nm, w_{slot2} = 350 nm, w_{Si} = 350 nm and h_{WG} = 250 nm. Special attention needs to be paid here to the different widths of slots, which is in accordance with the conformal mapping theory of a ring resonator [28]: the power of the outer slot of the whispering-gallery mode is larger than the inner slot, which results in the unbalance between the two slots of the DSHP ring. The unbalance will cause larger propagation loss (or lower Q factor) due to a large part of the light confined at the slot material-Ag interface compared to the DSHP waveguide without bending. In order to compensate such unbalance, the outer slot can be designed to be slightly wider than the inner one. Besides, the wider outer slot will also decrease the n_{eff} mismatch between the modes propagating in the open and closed areas of the DSHP ring, hence reducing the propagation loss of the designed, partly open DSHP ring resonator. Grating couplers are applied at each end of the device for the coupling between an optical fiber to on-chip devices.

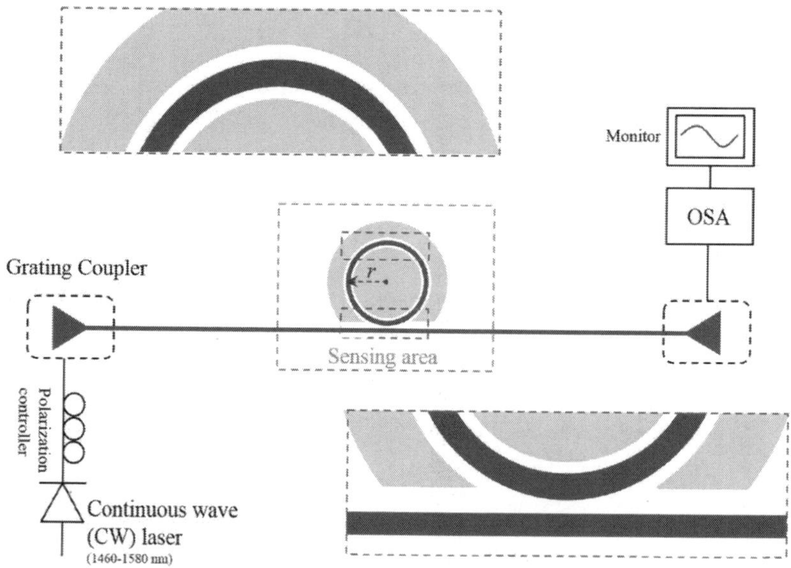

Figure 3. Schematic of the Si bus waveguide side-coupled double-slot hybrid plasmonic ring sensor. The sub-figures are the detailed structures of the coupling area and the DSHP ring. The measurement setup is also illustrated. OSA, optical spectrum analyzer.

3.2. Fabrication and Measurement Setup

The fabrication starts from a commercial SOI wafer with 250-nm crystalline Si on top of a 3 μm-thick SiO_2 buffer layer. E-beam lithography (EBL) is used to pattern the Si structure. Then, inductively-coupled plasma (ICP) dry etching with 10% over etch is performed, followed by a second EBL and ICP dry etching processes to fabricate the highly-efficient non-uniform grating couplers [29]. Finally, the pattern of the silver pads is introduced by the third E-beam exposure. After patterning, 20-nm Ge and 230-nm Ag layers are evaporated by a metal evaporation tool, where the Ge layer is used to increase the adhesive strength between silver and substrate material (SiO_2). Then, a metal lift-off process is used to open the silver structures.

The optical characterization is carried out by the grating coupler setup, as shown in Figure 3: the input/output optical fibers are placed above the grating couplers at each end of the tested device. The tunable continuous wave (CW) laser and polarization controller are connected to the input fiber, while the optical spectrum analyzer (OSA) is connected to the output fiber. The test liquids (100% and 80% concentrations of IPA in water) are directly dropped onto the surface of the tested sample. The transmission response of the tested device is then obtained after adjusting the positions of input and output optical fibers.

3.3. Characterization Results

The scanning electron microscope (SEM) view of the fabricated DSHP resonator is shown in Figure 4. The radius of the fabricated DSHP ring is about 6 μm, and the width of the Si ring and bus waveguide is 350 nm. The outer and inner slots have widths around 350 nm and 250 nm, respectively.

Figure 5 shows the transmission responses of the DSHP ring (black curves) and pure Si ring resonator (red curves) infiltrated with 100% IPA (solid curve) and 80% IPA (dashed curve). The green reference level is the transmission response of a straight Si waveguide with input/output gratings. For the DSHP ring resonator, the insertion loss, excluding grating coupler loss, is around 4 dB. The extinction ratio is larger than −25 dB, and the free-spectral-range (FSR) is about 14 nm. The wavelength shift of the DSHP ring is 3.3 nm, which is about 5-times higher than that of the Si ring (0.63 nm). The refractive index of 80% IPA at a 1550-nm wavelength can be estimated by the reference Si ring transmission responses: according to the simulation results, the Si ring with a 350-nm width has a sensitivity of 132.41 nm/RIU. Calculated from the resonance peak shift (0.63 nm), the refractive index change of the infiltrating material is 0.0048, which gives the refractive index of 80% IPA in water of 1.3691. Based on the data measured with a wavelength of 589 nm in [30], this value is reasonable. The corresponding sensitivity of the fabricated DSHP ring is 687.5 nm/RIU, slightly higher than the simulation result (~600 nm/RIU), which is caused by: (1) the coupling area providing additional sensitivity to the final results; and (2) the fabrication roughness of the Si and Ag elements inducing larger radiation losses compared to the theoretical situation and the radiated optical energy increasing the optical energy located in the infiltrating analytes, hence increasing the sensitivity.

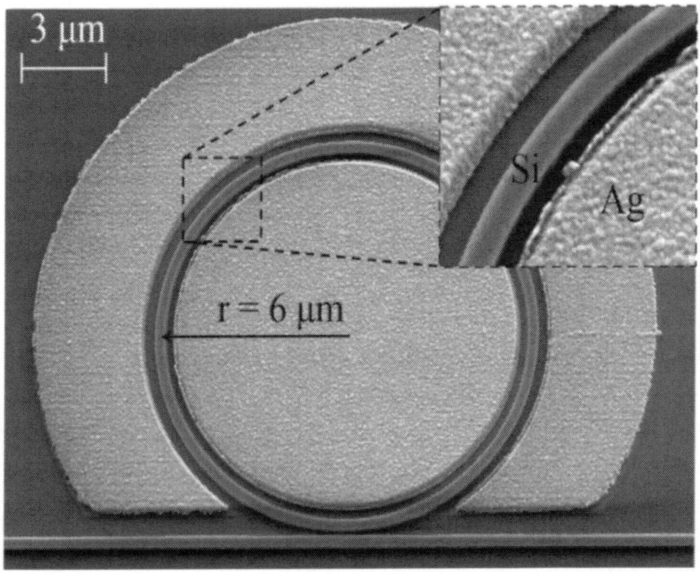

Figure 4. Scanning electron microscope picture of the fabricated double-slot hybrid plasmonic ring sensor.

Besides, compared to the theoretical Q_{abs} factor (larger than 10,000), the measured loaded Q factor is much lower (~300). Neglecting the radiation losses from waveguide bends, the coupling loss from the Si bus waveguide to the DSHP ring and fabrication-induced roughness, the phase mismatch between the open and closed areas of the DSHP ring resonator is the major reason that leads to the small loaded Q factor. It can be modified by decreasing the difference of n_{eff} between the two propagating modes, which will be shown in the following section.

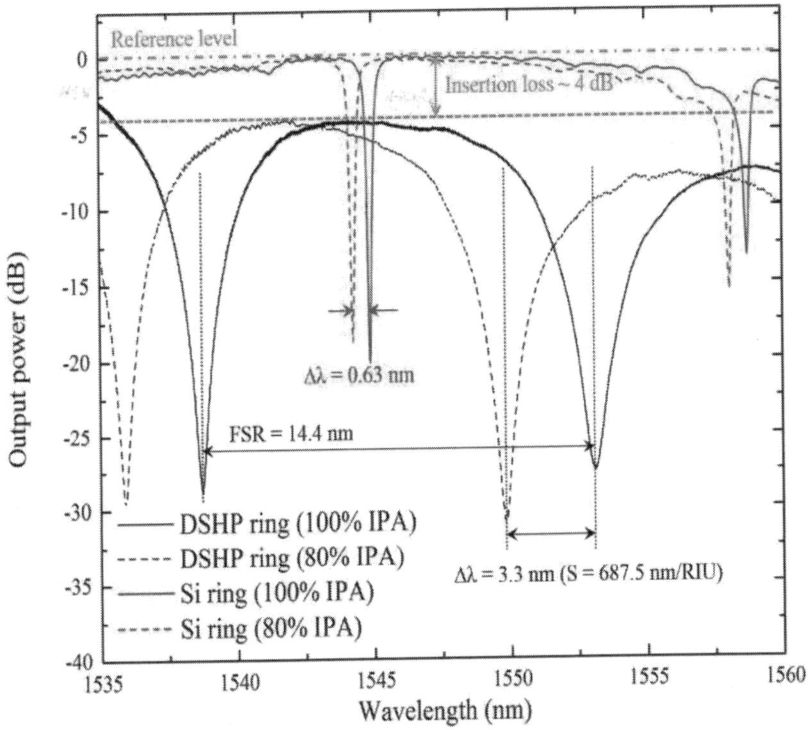

Figure 5. Characterization results of the double-slot hybrid plasmonic ring sensor (black curves) and a silicon ring resonator with the same radius (red curves) infiltrated with 100% and 80% 2-isopropanol. The reference level is the transmission response of the straight waveguide with input/output grating couplers.

4. LOADED *Q* FACTOR MODIFICATION

From the characterization results of the fabricated DSHP ring resonator, the loaded Q factor is much lower compared to the initial Q_{abs} factor shown in Figure 2b, which is mainly due to the phase mismatch between the different propagation modes (open and closed areas of the DSHP ring), in other words, the mismatch of n_{eff}. In order to compensate for such a mismatch, the difference between the n_{eff} of these two areas needs to be reduced. Since the n_{eff} of the closed DSHP ring is higher than that of the partly open area with the same w_{Si} due to better optical confinement, one can broaden the Si waveguide in the coupling area to get a higher Q factor, as shown in Figure 6a.

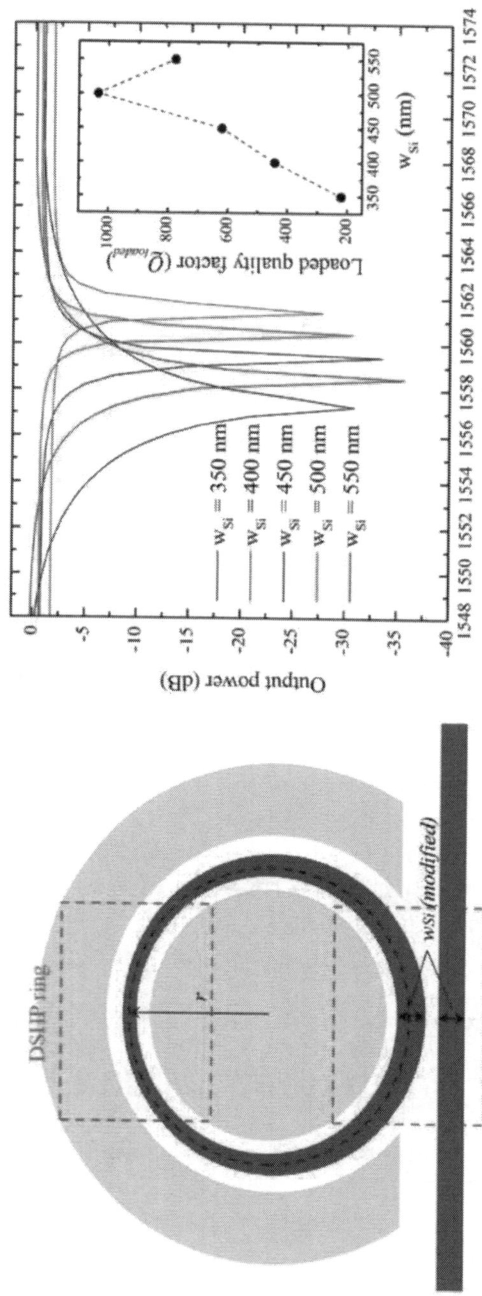

Figure 6. (**a**) Schematic of the modified double-slot hybrid plasmonic ring sensor. (**b**) Finite difference time domain (FDTD) simulation results of the modified DSHP ring sensor with different w_{Si}(*modified*). The sub-figure shows the loaded Q factor as a function of w_{Si}(*modified*).

In the coupling area, the widths of the Si bus waveguide and the partly open DSHP ring are modified, denoted byw_{Si}(*modified*). One needs to notice that the widths of the Si bus waveguide and the Si ridge of the open area of the DSHP ring have to be identical to achieve the phase match condition. The size and other parameters of the modified DSHP ring are similar to the fabricated one (w_{Si} = 350 nm, w_{slot1} = 250 nm and w_{slot2} = 350 nm). In order to decrease the coupling loss, Si tapers are used to connect the coupling area to the DSHP ring area.

The simulation is done using the finite difference time domain (FDTD) method, where the geometry of the DSHP ring is fixed, and w_{Si}(*modified*) increases from 350 nm to 550 nm. The transmission responses are shown in Figure 6b, where the different colors represent different w_{Si}(*modified*). One can see that when w_{Si}(*modified*) is around 500 nm, a maximum loaded$Q$ factor is achieved (larger than 1000), which means that the n_{eff} of the DSHP ring and the coupling are nearly matched, and the coupling loss comes to the minimum value.

5. DISCUSSION

For optical homogenous sensing applications, to measure the refractive index of liquids, the DSHP ring resonator shows high sensitivity to the infiltrating materials, giving a value as high as 687.5 nm/RIU as obtained experimentally. To further evaluate the performance of the DSHP ring sensor, a figure of merit (FOM) [31], defined by:

$$FOM = \frac{S}{\Delta\lambda} = \frac{S}{\lambda} Q \tag{10}$$

expressing the number of linewidth shifts for an infiltrating material refractive index change of one unit, can be applied. From the experimental results, the *FOM* of the DSHP ring sensor is around 97.5, which can be increased to 466 after loaded$Q$ factor modifications. The detection limit (*DL*), describing the smallest refractive index change that can be detected by the system, can be calculated by:

$$DL = \frac{f}{FOM} \tag{11}$$

where f is a factor expressing the fraction of the resonance linewidth that can be resolved. The f factor is highly dependent on the performance of the detection system used. From [16], we can estimate this value as 1/400 to evaluate the property of the DSHP ring. The detection limit of the fabricated DSHP ring is around 2.57×10^{-5} RIU (5.37×10^{-6} RIU after loaded Qfactor modification). Table 1 shows the comparisons between the fabricated DSHP ring sensor and other experimentally-demonstrated label-free ring sensors. Due to the high sensitivity of the DSHP ring, the *DL* after loaded Q factor modifications is low.

Another promising application of the DSHP ring resonator is as electro-optic modulator. The high-sensitivity to material filling the slots will result in the high efficiency of modulating optical devices. Moreover, the overlap between optical and electronic signals leads to a less complex fabrication process (no extra electrodes are required), where the electric field can be large due to the narrow gap between two plasmonic sheets. The performance of an EO polymer-based DSHP ring resonator modulator can be calculated by employing a typical high-quality EO polymer with properties as n = 1.5 and r_{33} = 300 pm/V [32], and assuming 80% of the resonant peak shift is provided by the slot material. Based on the given experimental results, the modulation voltage,

$$V = \frac{2\Delta n w_{slot}}{n^3 r_{33}} = \frac{2 w_{slot}}{FOM n^3 r_{33}} \tag{12}$$

for a 3-dB band shift is around 24 V. After loaded Q factor modification, it can decrease to as low a value as 5 V. The capacitance is estimated to be 0.245 fF, which is calculated from:

$$C = \varepsilon_0 \varepsilon_{EOP} \frac{2\pi r \cdot h_{WG}}{w} \tag{13}$$

where $2\pi r \cdot h_{WG}$ is the average area of the capacitance formed by inner and outer Ag rings and w is the total width of the DSHP ring ($w = w_{Si} + w_{slot1} + w_{slot2}$). By taking a reasonable resistance of electronic circuits (below 1500 Ω from [12]), the resistor-capacitor (RC) time constant is above 2 THz. For photon lifetime limitation, estimated from the Q factor, the bandwidth can reach a value larger than 600 GHz. The power consumption, expressing the energy dissipation during the charge-discharge cycles, is given by:

Table 1. Comparison between different kinds of ring-type label-free sensors (* results after loaded Q factor modifications).

Waveguide structure	**Width × Height**	**Radius**	**Q Factor**	**S (nm/RIU)**	**DL (RIU)**	**Reference**
Si waveguide	500 nm × 220 nm	≈ 5 μm	20,000	70	3.75×10^{-6}	[2]
Si slot waveguide	640 nm × 220 nm	> 5 μm	500	298	4.2×10^{-5}	[16]
SiN slot waveguide	1150 nm × 250	70 μm	1,800	212.13	2.3×10^{-5}	[15]
DSHP waveguide	950 nm × 250 nm	6 μm	300 (1034) *	687.5	2.57×10^{-5} (5.37×10^{-6}) *	This paper

$$W = \frac{1}{2} C V^2 f \tag{14}$$

According to the results given above, the power consumption is about 13.8 mW for 100-GHz modulation, which can decrease to 0.43 mW after loaded Q factor modification. The corresponding average power consumptions are 138 fJ/bit and 4.3 fJ/bit, respectively.

6. CONCLUSIONS

In this paper, we have experimentally demonstrated a DSHP ring resonator with ultra-high sensitivity to infiltrating materials, which shows the potential to be applied in homogeneous sensors and electro-optic polymer-based modulators.

Axisymmetric FEM simulations show that the Q factors of DSHP rings tend to their Q_{abs} factors when the radius is large enough (larger than about 6 μm), where the waveguide absorption loss dominants the performance of the DSHP ring rather than radiation loss. Then, optimizations are made for Q_{abs} factors by defining the q-parameter ($q = w_{Si}/w$) as the occupancy of the photonics mode. Simulation results show that the Q_{abs} factors for hybrid plasmonic-photonics mode ($0 < q < 1$) are much larger than that for a pure plasmonic slot waveguide ($q = 0$ or 1), as shown in Figure 2b, where the optimized Q_{abs}factors can reach as high a value as 300,000 in theory ($w = 1000$ nm and $q = 0.55$). Due to the interactions of the hybrid photonic-plasmonic mode, optimal q-parameters for sensitivity are also observed, which can reach 700 nm/RIU ($w = 500$ nm and $q = 0.35$), as shown in Figure 2d.

Based on simulation results, a DSHP ring resonator with a Si access waveguide by opening a coupling area (partly without the outer Ag ring) is designed, whose fabrication process is compatible with conventional SOI technology. Experimental results show that the sensitivity to the refractive index change of infiltrating materials (100% and 80% aqueous solution of IPA) of the fabricated DSHP can reach as high a value as 687.5 nm/RIU. The sensitivity is about five-times larger than that of Si ring resonator without plasmonic materials (132.41 nm/RIU). The measured sensitivity is slightly higher than simulation results (~600 nm/RIU), which is due to the sensitive coupling area and fabrication-induced roughness. Besides, the loaded Q factor of the DSHP ring is low (~300), which is mainly caused by the coupling loss between open and closed areas of the ring. In order to compensate for the coupling loss, as shown in Figure 6a, one can broaden the width of the Si ridge at the coupling area to decrease the difference of n_{eff} between open and closed areas of the DSHP ring. FDTD simulation results show that the modified *Q* factor can be over 1000, when phase matching is achieved, as shown in Figure 6b.

Compared to other experimentally-demonstrated types of optical ring sensors (Table 1), the fabricated DSHP ring has higher sensitivity and quite a low detection limit, which can be applied in advanced optical homogeneous sensors. Moreover, the special structure of the DSHP ring resonator is also promising for high-speed electro-optic modulator applications. Two plasmonic sheets can be used as electrodes with very low capacitance (~0.245 fF), and the narrow slot (less than 1 μm) can provide quite a large electric field. Taking the reasonable parameters of the electro-optic polymer, the applied voltage for a 3-dB band shift is about 5 V after loaded *Q* factor optimization, and the average energy consumption can reach as low value as 4.3 fJ/bit.

ACKNOWLEDGMENTS

This work was supported by "the Swedish Research Council (VR) through its Linnæus Center of Excellence ADOPT". Xu Sun acknowledges China Scholarship Council (CSC) for the financial support.

AUTHOR CONTRIBUTIONS

Xu Sun has done the design, performed the experiments and wrote the first manuscript; all the authors have discussed the problems and formulated the content.

CONFLICTS OF INTEREST

The authors declare no conflict of interest.

REFERENCES

1. Soref, R.A. Silicon-based optoelectronics. *Proc. IEEE* **1993**, *81*, 1687–1706.

2. De Vos, K.; Bartolozzi, I.; Schacht, E.; Bienstman, P.; Baets, R. Silicon-on-insulator microring resonator for sensitive and label-free biosensing. *Opt. Express* **2007**, *15*, 7610–7615.

3. Densmore, A.; Xu, D.-X.; Waldron, P.; Janz, S.; Cheben, P.; Lapointe, J.; Delâge, A.; Lamontagne, B.; Schmid, J.; Post, E. A silicon-on-insulator photonic wire based evanescent field sensor. *IEEE Photon. Technol. Lett.* **2006**, *18*, 2520–2522.

4. Dong, P.; Liao, S.; Feng, D.; Liang, H.; Zheng, D.; Shafiiha, R.; Kung, C.-C.; Qian, W.; Li, G.; Zheng, X. Low Vpp, ultralow-energy, compact, high-speed silicon electro-optic modulator. *Opt. Express* **2009**, *17*, 22484–22490.

5. Wangüemert-Pérez, G.J.; Cheben, P.; Ortega-Moñux, A.; Alonso-Ramos, C.; Pérez-Galacho, D.; Halir, R.; Molina-Fernández, I.; Xu, D.-X.; Schmid, J.H. Evanescent field waveguide sensing with subwavelength grating structures in silicon-on-insulator. *Opt. Lett.* **2014**, *39*, 4442–4445.

6. Jin, L.; Li, M.; He, J.-J. Highly-sensitive silicon-on-insulator sensor based on two cascaded micro-ring resonators with vernier effect. *Opt. Commun.* **2011**, *284*, 156–159.

7. Li, G.; Zheng, X.; Yao, J.; Thacker, H.; Shubin, I.; Luo, Y.; Raj, K.; Cunningham, J.E.; Krishnamoorthy, A.V. 25 Gb/s 1V-driving CMOS ring modulator with integrated thermal tuning. *Opt. Express* **2011**, *19*, 20435–20443.

8. Li, H.; Fan, X. Characterization of sensing capability of optofluidic ring resonator biosensors. *Appl. Phys. Lett.* **2010**, *97*.

9. Liu, A.; Jones, R.; Liao, L.; Samara-Rubio, D.; Rubin, D.; Cohen, O.; Nicolaescu, R.; Paniccia, M. A high-speed silicon optical modulator based on a metal–oxide–semiconductor capacitor. *Nature* **2004**, *427*, 615–618.

10. Wang, J.; Dai, D. Highly sensitive si nanowire-based optical sensor using a Mach–Zehnder interferometer coupled microring. *Opt. Lett.* **2010**, *35*, 4229–4231.

11. Xu, Q.; Manipatruni, S.; Schmidt, B.; Shakya, J.; Lipson, M. 12.5 gbit/s carrier-injection-based silicon micro-ring silicon modulators. *Opt. Express* **2007**, *15*, 430–436.

12. Alloatti, L.; Korn, D.; Palmer, R.; Hillerkuss, D.; Li, J.; Barklund, A.; Dinu, R.; Wieland, J.; Fournier, M.; Fedeli, J. 42.7 gbit/s electro-optic modulator in silicon technology. *Opt. Express* **2011**, *19*, 11841–11851.

13. Yamada, H.; Chu, T.; Ishida, S.; Arakawa, Y. Si photonic wire waveguide devices. *IEEE J. Sel. Top. Quantum Electron.***2006**, *12*, 1371–1379.

14. Schmidt, H.; Yin, D.; Barber, J.P.; Hawkins, A.R. Hollow-core waveguides and 2-d waveguide arrays for integrated optics of gases and liquids. *IEEE J. Sel. Top. Quantum Electron.* **2005**, *11*, 519–527.

15. Barrios, C.A.; Gylfason, K.B.; Sánchez, B.; Griol, A.; Sohlström, H.; Holgado, M.; Casquel, R. Slot-waveguide biochemical sensor. *Opt. Lett.* **2007**, *32*, 3080–3082.

16. Claes, T.; Molera, J.G.; de Vos, K.; Schacht, E.; Baets, R.; Bienstman, P. Label-free biosensing with a slot-waveguide-based ring resonator in silicon on insulator. *IEEE Photon. J.* **2009**, *1*, 197–204.

17. Liu, Q.; Tu, X.; Kim, K.W.; Kee, J.S.; Shin, Y.; Han, K.; Yoon, Y.-J.; Lo, G.-Q.; Park, M.K. Highly sensitive Mach–Zehnder interferometer biosensor based on silicon nitride slot waveguide. *Sens. Actuators B Chem.* **2013**, *188*, 681–688.

18. Prieto, F.; Sepulveda, B.; Calle, A.; Llobera, A.; Domínguez, C.; Abad, A.; Montoya, A.; Lechuga, L.M. An integrated optical interferometric nanodevice based on silicon technology for biosensor applications. *Nanotechnology* **2003**, *14*, 907–912.

19. Wang, X.; Grist, S.; Flueckiger, J.; Jaeger, N.A.; Chrostowski, L. Silicon photonic slot waveguide Bragg gratings and resonators. *Opt. Express* **2013**, *21*, 19029–19039.

20. Melikyan, A.; Alloatti, L.; Muslija, A.; Hillerkuss, D.; Schindler, P.; Li, J.; Palmer, R.; Korn, D.; Muehlbrandt, S.; van Thourhout, D. High-speed plasmonic phase modulators. *Nat. Photon.* **2014**, *8*, 229–233.

21. Dai, D.; He, S. Low-loss hybrid plasmonic waveguide with double low-index nano-slots. *Opt. Express* **2010**, *18*, 17958–17966.

22. Sun, X.; Dai, D.; Thylén, L.; Wosinski, L. High-sensitivity liquid refractive-index sensor based on a Mach-Zehnder interferometer with a double-slot hybrid plasmonic waveguide. *Opt. Express* **2015**, *23*, 25688–25699.

23. Oxborrow, M. Traceable 2-D finite-element simulation of the whispering-gallery modes of axisymmetric electromagnetic resonators. *IEEE Trans. Microw. Theory Tech.* **2007**, *55*, 1209–1218.

24. Wei, T.; Han, Y.; Li, Y.; Tsai, H.-L.; Xiao, H. Temperature-insensitive miniaturized fiber inline Fabry-Perot interferometer for highly sensitive refractive index measurement. *Opt. Express* **2008**, *16*, 5764–5769.

25. Bohren, C.F.; Huffman, D.R. *Absorption and scattering of light by small particles*; John Wiley & Sons: Weinheim, Germany, 2008.

26. Sun, X.; Wosinski, L.; Thylen, L. Nanoscale surface plasmon polariton disk resonators, a theoretical analysis. *IEEE J. Sel. Top. Quantum Electron.* **2016**, *22*.

27. Lou, F.; Dai, D.; Thylen, L.; Wosinski, L. Design and analysis of ultra-compact EO polymer modulators based on hybrid plasmonic microring resonators. *Opt. Express* **2013**, *21*, 20041–20051.

28. Saleh, B.E.; Teich, M.C. *Resonator Optics in Fundamentals of Photonics*; John Wiley & Sons: New York, NY, USA, 1991.

29. Tang, Y.; Wang, Z.; Wosinski, L.; Westergren, U.; He, S. Highly efficient nonuniform grating coupler for silicon-on-insulator nanophotonic circuits. *Opt. Lett.* **2010**, *35*, 1290–1292.

30. Lide, D.R. *CRC Handbook of Chemistry and Physics*; CRC Press: Boca Raton, FL, USA, 2004.

31. Sherry, L.J.; Jin, R.; Mirkin, C.A.; Schatz, G.C.; van Duyne, R.P. Localized surface plasmon resonance spectroscopy of single silver triangular nanoprisms. *Nano Lett.* **2006**, *6*, 2060–2065.

32. Dalton, L.; Robinson, B.; Jen, A.; Ried, P.; Eichinger, B.; Sullivan, P.; Akelaitis, A.; Bale, D.; Haller, M.; Luo, J. Electro-optic coefficients of 500 pm/v and beyond for organic materials. *Proc. SPIE* **2005**, *5935*.

INDEX